Managing health and safety in construction

Construction (Design and Management) Regulations 2015

Guidance on Regulations

London: TSO

Published by TSO (The Stationery Office), part of Williams Lea, and available from:

Online
www.tsoshop.co.uk

Mail, Telephone & E-mail
TSO
PO Box 29, Norwich, NR3 1GN
Telephone orders/General enquiries: 0333 202 5070
E-mail: customer.services@tso.co.uk
Textphone: 0333 202 5077

Published for the Health and Safety Executive under licence from the Controller of His Majesty's Stationery Office.

© Crown copyright 2015

First published 2015

ISBN 9780717666263

SD000273 5/25

This information is licensed under the Open Government Licence v3.0. To view this licence, visit http://www.nationalarchives.gov.uk/doc/open-government-licence/ OGL

Any enquiries regarding this publication should be sent to: HSE.Online@hse.gov.uk

Some images and illustrations in this publication may not be owned by the Crown and cannot be reproduced without permission of the copyright owner. Where we have identified any third-party copyright information you will need to obtain permission from the copyright holders concerned. Enquiries should be sent to HSE.Online@hse.gov.uk

This guidance is issued by the Health and Safety Executive. Following the guidance is not compulsory, unless specifically stated, and you are free to take other action. But if you do follow the guidance you will normally be doing enough to comply with the law. Health and safety inspectors seek to secure compliance with the law and may refer to this guidance

Printed in the United Kingdom for The Stationery Office.

Contents

Introduction		5
PART 1	Commencement, interpretation and application	10
Regulation 1	Citation and commencement	10
Regulation 2	Interpretation	10
Regulation 3	Application in and outside Great Britain	13
PART 2	Client duties	14
Regulation 4	Client duties in relation to managing projects	14
Regulation 5	Appointment of the principal designer and the principal contractor	15
Regulation 6	Notification	20
Regulation 7	Application to domestic clients	21
PART 3	Health and safety duties and roles	23
Regulation 8	General duties	23
Regulation 9	Duties of designers	26
Regulation 10	Designs prepared or modified outside Great Britain	26
Regulation 11	Duties of a principal designer in relation to health and safety at the pre-construction phase	30
Regulation 12	Construction phase plan and health and safety file	34
Regulation 13	Duties of a principal contractor in relation to health and safety at the construction phase	36
Regulation 14	Principal contractor's duties to consult and engage with workers	42
Regulation 15	Duties of contractors	43
PART 4	General requirements for all construction sites	50
Regulation 16	Application of Part 4	50
Regulation 17	Safe places of construction work	50
Regulation 18	Good order and site security	50
Regulation 19	Stability of structures	51
Regulation 20	Demolition or dismantling	51
Regulation 21	Explosives	51
Regulation 22	Excavations	52
Regulation 23	Cofferdams and caissons	52
Regulation 24	Reports of inspections	53
Regulation 25	Energy distribution installations	54
Regulation 26	Prevention of drowning	54

Regulation 27	Traffic routes	54
Regulation 28	Vehicles	55
Regulation 29	Prevention of risk from fire, flooding or asphyxiation	56
Regulation 30	Emergency procedures	56
Regulation 31	Emergency routes and exits	56
Regulation 32	Fire detection and fire-fighting	57
Regulation 33	Fresh air	57
Regulation 34	Temperature and weather protection	57
Regulation 35	Lighting	58
PART 5	General	59
Regulation 36	Enforcement in respect of fire	59
Regulation 37	Transitional and saving provisions	59
Regulation 38	Revocation and consequential amendments	60
Regulation 39	Review	60
SCHEDULE 1	Particulars to be notified under regulation 6	62
SCHEDULE 2	Minimum welfare facilities required for construction sites	63
SCHEDULE 3	Work involving particular risks	66
SCHEDULE 4	Transitional and saving provisions	67
SCHEDULE 5	Amendments	71
Appendix 1	The general principles of prevention	73
Appendix 2	Pre-construction information	74
Appendix 3	The construction phase plan	77
Appendix 4	The health and safety file	81
Appendix 5	How different types of information relate to and influence each other in a construction project involving more than one contractor: A summary	84
Appendix 6	Working for a domestic client	85
References and further reading		87
Glossary of acronyms and terms		88
Further information		90

Introduction

About this book

1	This book gives guidance on the Construction (Design and Management) Regulations 2015 (CDM 2015). These Regulations cover the management of health, safety and welfare when carrying out construction projects. Subject to some transitional provisions (see paragraphs 181–186), CDM 2015 replaces the Construction (Design and Management) Regulations 2007 (CDM 2007) from 6 April 2015. From this date, the Approved Code of Practice (ACOP) which provides supporting guidance on CDM 2007 is withdrawn.

Who this book is for

2	This guidance is for people with legal duties under CDM 2015. It explains what they must or should do to comply with the law. Any actions taken should always be proportionate to the risks in the construction project.

3	Under CDM 2015, organisations or individuals can be one or more dutyholder for a project. The different dutyholders are summarised in Table 1. The table does not include all the duties, nor does it distinguish between duties that are absolute (dutyholders 'must' comply with them), and duties that are qualified by terms such as 'as far as practicable' or 'as far as reasonably practicable' (dutyholders 'should' comply with them). See the Glossary for a more detailed explanation of these terms. Guidance on specific regulations explains the duties under CDM 2015 in more detail.

Table 1 A summary of roles and duties under CDM 2015

CDM dutyholders:* Who are they?	Summary of role/main duties
Clients are organisations or individuals for whom a construction project is carried out.	Make suitable arrangements for managing a project. This includes making sure: ■ other dutyholders are appointed; ■ sufficient time and resources are allocated. Make sure: ■ relevant information is prepared and provided to other dutyholders; ■ the principal designer and principal contractor carry out their duties; ■ welfare facilities are provided. See paragraphs 23–52 for more guidance.
Domestic clients are people who have construction work carried out on their own home, or the home of a family member that is **not** done as part of a business, whether for profit or not.	Domestic clients are in scope of CDM 2015, but their duties as a client are normally transferred to: ■ the contractor, on a single contractor project; or; ■ the principal contractor, on a project involving more than one contractor. However, the domestic client can choose to have a written agreement with the principal designer to carry out the client duties. See paragraphs 53–56 for more guidance.
Designers are those, who as part of a business, prepare or modify designs for a building, product or system relating to construction work.	When preparing or modifying designs, to eliminate, reduce or control foreseeable risks that may arise during: ■ construction; and ■ the maintenance and use of a building once it is built. Provide information to other members of the project team to help them fulfil their duties. See paragraphs 72–93 for more guidance.
Principal designers** are designers appointed by the client in projects involving more than one contractor. They can be an organisation or an individual with sufficient knowledge, experience and ability to carry out the role.	Plan, manage, monitor and coordinate health and safety in the pre-construction phase of a project. This includes: ■ identifying, eliminating or controlling foreseeable risks; ■ ensuring designers carry out their duties. Prepare and provide relevant information to other dutyholders. Provide relevant information to the principal contractor to help them plan, manage, monitor and coordinate health and safety in the construction phase. See paragraphs 94–115 for more guidance.

	Principal contractors are contractors appointed by the client to coordinate the construction phase of a project where it involves more than one contractor.	Plan, manage, monitor and coordinate health and safety in the construction phase of a project. This includes: ■ liaising with the client and principal designer; ■ preparing the construction phase plan; ■ organising cooperation between contractors and coordinating their work. Ensure: ■ suitable site inductions are provided; ■ reasonable steps are taken to prevent unauthorised access; ■ workers are consulted and engaged in securing their health and safety; and ■ welfare facilities are provided. See paragraphs 110–146 for more guidance.
	Contractors are those who do the actual construction work and can be either an individual or a company.	Plan, manage and monitor construction work under their control so that it is carried out without risks to health and safety. For projects involving more than one contractor, coordinate their activities with others in the project team – in particular, comply with directions given to them by the principal designer or principal contractor. For single-contractor projects, prepare a construction phase plan. See paragraphs 147–179 for more guidance.
	Workers are the people who work for or under the control of contractors on a construction site.	They must: ■ be consulted about matters which affect their health, safety and welfare; ■ take care of their own health and safety and others who may be affected by their actions; ■ report anything they see which is likely to endanger either their own or others' health and safety; ■ cooperate with their employer, fellow workers, contractors and other dutyholders.

*Organisations or individuals can carry out the role of more than one dutyholder, provided they have the skills, knowledge, experience and (if an organisation) the organisational capability to carry out those roles in a way that secures health and safety.

** Principal designers are **not** a direct replacement for CDM co-ordinators. The range of duties they carry out is different to those undertaken by CDM co-ordinators under CDM 2007 (see paragraphs 181–186 for information about transitional arrangements).

Key elements to securing construction health and safety

4   The key elements, include:

(a)   managing the risks by applying the **general principles of prevention**;
(b)   **appointing** the right people and organisations at the right time;
(c)   making sure everyone has the **information, instruction, training and supervision** they need to carry out their jobs in a way that secures health and safety;
(d)   dutyholders **cooperating and communicating** with each other and **coordinating** their work; and
(e)   **consulting workers and engaging** with them to promote and develop effective measures to secure health, safety and welfare.

General principles of prevention

5   These set out the principles dutyholders should use in their approach to identifying the measures they should take to control the risks to health and safety in a particular project. The general principles of prevention are set out in full in Appendix 1, but in summary they are to:

(a)   avoid risks where possible;
(b)   evaluate those risks that cannot be avoided; and
(c)   put in place proportionate measures that control them at source.

CDM 2015 requires designers, principal designers, principal contractors and contractors to take account of the principles in carrying out their duties.

Appointing the right organisations and people at the right time

6   Appointing the right organisations and individuals to complete a particular project is fundamental to its success, including health and safety performance.

Appointing designers and contractors

7   Anyone responsible for appointing designers (including principal designers) or contractors (including principal contractors) to work on a project must ensure that those appointed have the skills, knowledge and experience to carry out the work in a way that secures health and safety. If those appointed are an organisation, they must also have the appropriate organisational capability. Those making the appointments must establish that those they appoint have these qualities **before** appointing them. Similarly, any designers or contractors seeking appointment as individuals must ensure they have the necessary skills, knowledge and experience.

8   Dutyholders should be appointed at the right time. For example, clients must appoint principal designers and principal contractors as soon as practicable and **before** the start of the construction phase, so they have enough time to carry out their duties to plan and manage the pre-construction and construction phases respectively. See paragraphs 35–40 and 58–65 for guidance on making these appointments and their timing.

Contractors appointing anyone for work on a construction site

9   When contractors appoint anyone to carry out work on a construction site, they must make sure that those they appoint have, or are in the process of gaining, the right skills, knowledge, training and experience (see paragraphs 162–168). Not everyone will have these qualities and, if they do not, appointments should be made on the basis that they are capable of gaining them.

Supervision, instructions and information

10   The level of supervision, instructions and information required will depend on the risks involved in the project and the level of skills, knowledge, training and experience of the workforce. Contractors (including principal contractors) must make sure supervision is effective and suitable site inductions are provided along with other information – such as the procedures to be followed in the event of serious and imminent danger to health and safety (see paragraphs 169–173).

Cooperating, communicating and coordinating

11   Dutyholders must cooperate with each other and coordinate their work to ensure health and safety. They must also communicate with each other to make sure everyone understands the risks and the measures to control those risks. For example, through regular dialogue between the client, the principal designer and principal contractor to ensure they have the time and resources to plan, manage, monitor and coordinate the pre-construction and construction phases (see paragraphs 66–67).

Consulting and engaging with workers

12   Workplaces where workers are consulted and engaged in decisions about health and safety measures are safer and healthier. Consultation about health and safety is two way. It involves giving information to workers, listening to them and taking account of what they say before decisions are made by the dutyholder. For example, hold meetings before work starts to discuss the work planned for the day, identify risks and agree appropriate control measures. Involving workers helps those responsible for health and safety to manage it in a practical way by:

(a)   helping spot workplace risks and knowing what to do about them;
(b)   making sure health and safety controls are appropriate;
(c)   increasing the level of commitment to working in a safe and healthy way.

13   Workers must be consulted in good time. The:

(a)   Safety Representatives and Safety Committees Regulations 1977; and
(b)   Health and Safety (Consultation with Employees) Regulations 1996

require employers to consult their workforce about health and safety (see *Consulting employees on health and safety: A brief guide to the law*[1] and www.hse.gov.uk/involvement/ for more information). In workplaces where a trade union is recognised, consultation should be through union health and safety representatives. In non-unionised workplaces, consultation should be either direct with workers or through other elected representatives.

14   In addition, CDM 2015 places a specific duty on principal contractors to consult and engage with workers (see paragraphs 143–146).

15   To help managers build worker engagement into the everyday running of their businesses and sites, HSE and the industry's Leadership and Worker Engagement Forum have produced the free online *Leadership and Worker Involvement Toolkit* (LWIT). It includes useful resources, such as case studies and videos and can be found at www.hse.gov.uk/construction.

Other regulations that apply to construction

16   Other health and safety regulations also apply to construction. For example, the Work at Height Regulations 2005 and the Control of Asbestos Regulations 2012 (see www.hse.gov.uk/construction for more information).

Presentation

17   The text of the Regulations is set out in *italics* and the accompanying guidance is in normal type. Some of the regulations are preceded by a short boxed summary of the main duties imposed by that regulation or regulations to help the reader navigate the document.

18   Where this guidance refers to a building, it means any structure as defined in regulation 2(1). Other terms and acronyms are explained in the Glossary.

PART 1 Commencement, interpretation and application

Regulation 1 Citation and commencement

Regulation 1

These Regulations may be cited as the Construction (Design and Management) Regulations 2015 and come into force on 6th April 2015 immediately after the Mines Regulations 2014.

Guidance 1

19 The Regulations come into force on 6 April 2015 and replace CDM 2007.

20 The Regulations are subject to certain transitional provisions which recognise there will be projects that started before CDM 2015 comes into force. Guidance on these is set out in paragraphs 181–186, and, in particular, recognises that any CDM co-ordinator appointed under CDM 2007 may continue to carry out the duties they had under CDM 2007 for up to six months after CDM 2015 comes into force. This will allow a principal designer to be appointed and replace the CDM co-ordinator within that period.

Regulations 2 and 3 define the terms used and the scope of CDM 2015.

Regulation 2 Interpretation

Regulation 2

(1) In these Regulations—

"the 1974 Act" means the Health and Safety at Work etc. Act 1974;

"the 2007 Regulations" means the Construction (Design and Management) Regulations 2007;

"the Management Regulations" means the Management of Health and Safety at Work Regulations 1999;

"business" means a trade, business or other undertaking (whether for profit or not);

"client" means any person for whom a project is carried out;

"construction phase" means the period of time beginning when construction work in a project starts and ending when construction work in that project is completed;

"construction phase plan" means a plan drawn up under regulations 12 or 15;

"construction site" includes any place where construction work is being carried out or to which the workers have access, but does not include a workplace within the site which is set aside for purposes other than construction work;

"construction work" means the carrying out of any building, civil engineering or engineering construction work and includes—

Regulation 2

(a) the construction, alteration, conversion, fitting out, commissioning, renovation, repair, upkeep, redecoration or other maintenance (including cleaning which involves the use of water or an abrasive at high pressure, or the use of corrosive or toxic substances), de-commissioning, demolition or dismantling of a structure;

(b) the preparation for an intended structure, including site clearance, exploration, investigation (but not site survey) and excavation (but not pre-construction archaeological investigations), and the clearance or preparation of the site or structure for use or occupation at its conclusion;

(c) the assembly on site of prefabricated elements to form a structure or the disassembly on site of the prefabricated elements which, immediately before such disassembly, formed a structure;

(d) the removal of a structure, or of any product or waste resulting from demolition or dismantling of a structure, or from disassembly of prefabricated elements which immediately before such disassembly formed such a structure;

(e) the installation, commissioning, maintenance, repair or removal of mechanical, electrical, gas, compressed air, hydraulic, telecommunications, computer or similar services which are normally fixed within or to a structure,

but does not include the exploration for, or extraction of, mineral resources, or preparatory activities carried out at a place where such exploration or extraction is carried out;

"contractor" means any person (including a non-domestic client) who, in the course or furtherance of a business, carries out, manages or controls construction work;

"design" includes drawings, design details, specifications and bills of quantities (including specification of articles or substances) relating to a structure, and calculations prepared for the purpose of a design;

"designer" means any person (including a client, contractor or other person referred to in these Regulations) who in the course or furtherance of a business—

(a) prepares or modifies a design; or
(b) arranges for, or instructs, any person under their control to do so,

relating to a structure, or to a product or mechanical or electrical system intended for a particular structure, and a person is deemed to prepare a design where a design is prepared by a person under their control;

"domestic client" means a client for whom a project is being carried out which is not in the course or furtherance of a business of that client;

"excavation" includes any earthwork, trench, well, shaft, tunnel or underground working;

"the general principles of prevention" means the general principles of prevention specified in Schedule 1 to the Management Regulations;

"health and safety file" means a file prepared under regulation 12(5);

"inspector for the Executive" means an inspector within the meaning given in section 53(1) of the 1974 Act;

Regulation 2

"loading bay" means any facility for loading or unloading;

"place of work" means any place which is used by any person at work for the purposes of construction work or for the purposes of any activity arising out of or in connection with construction work;

"pre-construction information" means information in the client's possession or which is reasonably obtainable by or on behalf of the client, which is relevant to the construction work and is of an appropriate level of detail and proportionate to the risks involved, including—

    (a)   information about—
        (i)   the project;
        (ii)   planning and management of the project;
        (iii)   health and safety hazards, including design and construction hazards and how they will be addressed; and
    (b)   information in any existing health and safety file;

"pre-construction phase" means any period of time during which design or preparatory work is carried out for a project and may continue during the construction phase;

"principal contractor" means the contractor appointed under regulation 5(1)(b) to perform specified duties in regulations 12 to 14;

"principal designer" means the designer appointed under regulation 5(1)(a) to perform specified duties in regulations 11 and 12;

"project" means a project which includes or is intended to include construction work and includes all planning, design, management or other work involved in a project until the end of the construction phase;

"site rules" means rules which are drawn up for a particular construction site and are necessary for health or safety purposes;

"structure" means—

    (a)   any building, timber, masonry, metal or reinforced concrete structure, railway line or siding, tramway line, dock, harbour, inland navigation, tunnel, shaft, bridge, viaduct, waterworks, reservoir, pipe or pipeline, cable, aqueduct, sewer, sewage works, gasholder, road, airfield, sea defence works, river works, drainage works, earthworks, lagoon, dam, wall, caisson, mast, tower, pylon, underground tank, earth retaining structure or structure designed to preserve or alter any natural feature, and fixed plant;
    (b)   any structure similar to anything specified in paragraph (a);
    (c)   any formwork, falsework, scaffold or other structure designed or used to provide support or means of access during construction work,

and any reference to a structure includes part of a structure;

"traffic route" means a route for pedestrian traffic or for vehicles and includes any doorway, gateway, loading bay or ramp;

"vehicle" includes any mobile work equipment;

"work equipment" means any machinery, appliance, apparatus, tool or installation for use at work (whether exclusively or not);

Regulation 2

"working day" means any day on which construction work takes place;

"workplace" means a workplace within the meaning of regulation 2(1) of the Workplace (Health, Safety and Welfare) Regulations 1992 other than a construction site.

(2) Any reference in these Regulations to a plan, rule, document, report or copy includes a copy or electronic version which is —

(a) capable of being retrieved or reproduced when required; and
(b) secure from loss or unauthorised interference.

Regulation 3 Application in and outside Great Britain

Regulation 3

These Regulations apply—

(a) in Great Britain; and
(b) to premises and activities outside Great Britain to which sections 1 to 59 and 80 to 82 of the 1974 Act apply by virtue of articles 9 and 11(1)(a) of the Health and Safety at Work etc. Act 1974 (Application outside Great Britain) Order 2013.

Guidance 3

21 CDM 2015 applies to all construction projects in Great Britain. Through the Health and Safety at Work etc Act 1974 (Application outside Great Britain) Order 2013, it also applies to construction work carried out in:

(a) the territorial sea (see Glossary); and
(b) connection with, or preparatory to, construction of any renewable energy structure in the renewable energy zone (see Glossary).

22 Other than Part 4, the Regulations apply to construction projects as a whole – that is, the whole construction process from concept to completion. Part 4 sets out a number of provisions that only relate to work carried out on construction sites.

PART 2 Client duties

> **Regulations 4 and 5** set out the client's duty to make suitable arrangements for managing a project and maintaining and reviewing these arrangements throughout, so the project is carried out in a way that manages the health and safety risks. For projects involving more than one contractor, these regulations require the client to appoint a principal designer and a principal contractor and make sure they carry out their duties.

Regulation 4 Client duties in relation to managing projects

Regulation 4

(1) A client must make suitable arrangements for managing a project, including the allocation of sufficient time and other resources.

(2) Arrangements are suitable if they ensure that—

(a) the construction work can be carried out, so far as is reasonably practicable, without risks to the health or safety of any person affected by the project; and
(b) the facilities required by Schedule 2 are provided in respect of any person carrying out construction work.

(3) A client must ensure that these arrangements are maintained and reviewed throughout the project.

(4) A client must provide pre-construction information as soon as is practicable to every designer and contractor appointed, or being considered for appointment, to the project.

(5) A client must ensure that—

(a) before the construction phase begins, a construction phase plan is drawn up by the contractor if there is only one contractor, or by the principal contractor; and
(b) the principal designer prepares a health and safety file for the project, which—
    (i) complies with the requirements of regulation 12(5);
    (ii) is revised from time to time as appropriate to incorporate any relevant new information; and
    (iii) is kept available for inspection by any person who may need it to comply with the relevant legal requirements.

(6) A client must take reasonable steps to ensure that—

(a) the principal designer complies with any other principal designer duties in regulations 11 and 12; and

| Regulation 4 |

(b) the principal contractor complies with any other principal contractor duties in regulations 12 to 14;

(7) If a client disposes of the client's interest in the structure, the client complies with the duty in paragraph (5)(b)(iii) by providing the health and safety file to the person who acquires the client's interest in the structure and ensuring that that person is aware of the nature and purpose of the file.

(8) Where there is more than one client in relation to a project—

(a) one or more of the clients may agree in writing to be treated for the purposes of these Regulations as the only client or clients; and
(b) except for the duties specified in sub-paragraph (c) only the client or clients agreed in paragraph (a) are subject to the duties owed by a client under these Regulations;
(c) the duties in the following provisions are owed by all clients—
   (i) regulation 8(4); and
   (ii) paragraph (4) and regulation 8(6) to the extent that those duties relate to information in the possession of the client.

Regulation 5 Appointment of the principal designer and the principal contractor

| Regulation 5 |

(1) Where there is more than one contractor, or if it is reasonably foreseeable that more than one contractor will be working on a project at any time, the client must appoint in writing—

(a) a designer with control over the pre-construction phase as principal designer; and
(b) a contractor as principal contractor.

(2) The appointments must be made as soon as is practicable, and in any event, before the construction phase begins.

(3) If the client fails to appoint a principal designer, the client must fulfil the duties of the principal designer in regulations 11 and 12.

(4) If the client fails to appoint a principal contractor, the client must fulfil the duties of the principal contractor in regulations 12 to 14.

| Guidance 4, 5 |

Who is a client?

23 CDM 2015 defines a client as anyone for whom a construction project is carried out (see regulation 2(1)). This definition includes both non-domestic (or **'commercial'**) clients and **'domestic'** clients (ie clients for whom a construction project is carried out which is **not** done in connection with a business). The Regulations apply in full to commercial clients, but for domestic clients, the effect of regulation 7 is to pass the client duties on to other dutyholders. This includes the principal designer and principal contractor duties falling to the designer and contractor in control of the pre-construction and construction phases, where the domestic client does not make these appointments. Further guidance on how the Regulations apply to domestic clients is set out in paragraphs 53–56.

24 The guidance in paragraphs 25–52 applies to commercial clients, and any reference to 'clients' elsewhere in this book should be read as referring to

Guidance 4, 5

commercial clients **only,** unless specific reference to domestic clients is made. Commercial clients are organisations or individuals for whom a construction project is carried out in connection with a business, whether the business operates for profit or not. This includes clients based overseas who commission construction projects in Great Britain.

25   In any project there may be more than one client. Regulation 4(8) may be used where there is more than one client, but all agree that only one of them should be responsible for carrying out the requirements of CDM 2015.

26   In some circumstances, it may not be clear who the client or clients are. Any uncertainty should be resolved as early as possible by considering who:

(a)   ultimately decides what is to be constructed, where, when and by whom;
(b)   commissions the design and construction work (the employer in contract terminology);
(c)   initiates the work;
(d)   is at the head of the procurement chain; and
(e)   appoints contractors (including the principal contractor) and designers (including the principal designer).

If there is still doubt about who the client or clients are, all the possible clients should agree that one or more of them is treated as the client for the purposes of CDM 2015. It is in the interests of all those involved to identify and agree who the client or clients are. If not, they may all be considered to have client duties under CDM 2015.

27   Those clients who have not been identified as the client for the purposes of CDM 2015 will still have duties. These are to:

(a)   provide any information in their possession that may be relevant to help pull together the pre-construction information; and
(b)   cooperate with anyone involved in the project.

Why is the client important?

28   The client has a major influence over the way a project is procured and managed. Regardless of the size of the project, the client has contractual control, appoints designers and contractors, and determines the money, time and other resources available.

What must a client do?

29   CDM 2015 makes the client accountable for the impact their decisions and approach have on health, safety and welfare on the project. This section provides guidance on client duties under regulations 4 and 5. See guidance on the other duties a client has under regulation 6 (Notification – paragraphs 47–52) and regulation 8 (General duties – paragraphs 57–71).

Making suitable arrangements for managing a project
30   Most clients, particularly those who only occasionally commission construction work, will not be experts in the construction process. For this reason, they are not required to take an active role in managing the work. However, the client is required to make suitable arrangements for managing the project so that health, safety and welfare is secured.

Guidance   **4, 5**

31   To be suitable, the arrangements should focus on the needs of the particular project and be proportionate to the size of the project and risks involved in the work. Arrangements should include:

(a)   assembling the project team – appointing designers (including a principal designer) and contractors (including a principal contractor). See paragraphs 35–40 for more guidance;
(b)   ensuring the roles, functions and responsibilities of the project team are clear;
(c)   ensuring sufficient resources and time are allocated for each stage of the project – from concept to completion;
(d)   ensuring effective mechanisms are in place for members of the project team to communicate and cooperate with each other and coordinate their activities;
(e)   how the client will take reasonable steps to ensure that the principal designer and principal contractor comply with their separate duties. This could take place at project progress meetings or via written updates;
(f)   setting out the means to ensure that the health and safety performance of designers and contractors is maintained throughout;
(g)   ensuring that workers are provided with suitable welfare facilities for the duration of construction work.

32   Clients should take ownership of these arrangements and ensure they communicate them clearly to other dutyholders. Clients could prepare a clear 'client's brief' as a way of setting out the arrangements. The client brief normally:

(a)   sets out the main function and operational requirements of the finished project;
(b)   outlines how the project is expected to be managed including its health and safety risks;
(c)   sets a realistic timeframe and budget; and
(d)   covers other relevant matters, such as establishing design direction and a single point of contact in the client's organisation.

33   Where the range and nature of risks involved in the work warrants it, the management arrangements should also include:

(a)   the expected standards of health and safety, including safe working practices, and the means by which these standards will be maintained throughout;
(b)   what is expected from the design team in terms of the steps they should reasonably take to ensure their designs help manage foreseeable risks during the construction phase and when maintaining and using the building once it is built;
(c)   the arrangements for commissioning the new building and a well-planned handover procedure to the new user.

34   If a client needs help in making these arrangements, the principal designer should be in a position to help with this. Clients could also draw on the advice of a competent person if they are required to appoint such a person under the Management of Health and Safety at Work Regulations 1999 ('the Management Regulations' – see www.hse.gov.uk/toolbox/managing/managingtherisks.htm for more information).

Assembling the project team

35   The management arrangements must cover what clients will do to ensure that the people and organisations they appoint have the skills, knowledge, experience and (if an organisation) the organisational capability to manage health and safety risks (see paragraphs 58–65 for further guidance). This applies to both:

(a)   single contractor projects where the client will appoint the contractor and/or designers directly; and

Guidance 4, 5

(b) projects involving more than one contractor where the client is required to appoint, in writing, a principal designer and a principal contractor.

36 The extent of the checks a client must make into the capabilities of dutyholders they appoint will depend on the complexity of the project and the range and nature of the risks involved. See paragraphs 58–62 for further guidance on the help available to clients in selecting the right dutyholder.

Appointing principal designers and principal contractors

37 The principal designer should be appointed as early as possible in the design process, if practicable at the concept stage. Appointing the principal designer early will provide the client with help in matters such as pulling together the pre-construction information (see paragraphs 42–43) and giving the principal designer enough time to carry out their duties. The duration of the principal designer's appointment should take into account any design work which may continue into the construction phase or any issues that may arise during construction involving the need to make suitable modifications to the designs. For projects involving early work by a concept architect or project management company where a design and build contractor or novated designer is subsequently involved, it may be appropriate for the initial principal designer appointment to be ended and a new principal designer appointed.

38 The principal contractor should be appointed early enough in the pre-construction phase to help the client meet their duty to ensure a construction phase plan is drawn up before the construction phase starts. This also gives the principal contractor time to carry out their duties, such as preparing the construction phase plan and liaising with the principal designer in sharing any relevant information for health and safety.

39 The principal designer should be in place for as long as there is a need for their role to be performed. But where a principal designer's appointment finishes before the end of the project, the client should ensure that the principal contractor is fully briefed on matters arising from designs relevant to any subsequent construction work. The client should also make sure that the principal designer passes the health and safety file to the principal contractor so it can be revised during the remainder of the project if necessary.

40 If a client fails to appoint either a principal designer or a principal contractor, the client must carry out their duties.

Maintaining and reviewing the management arrangements

41 The client must maintain and review their arrangements to ensure they remain relevant throughout the life of the project. Some projects do not go smoothly and clients may experience difficulties and delays as they progress. Examples of actions the client can take to maintain and review their arrangements are:

(a) establishing key milestones so they can assess the progress of the project and determine whether health and safety standards are being met;
(b) where necessary, seeking advice. On larger projects, the client may value an independent review of standards;
(c) ensuring arrangements for handing over the building to a new user are sufficient to protect anyone (including members of the public) who may be affected by risks arising from any ongoing construction work, eg snagging work.

Pre-construction information

42 Pre-construction information is information already in the client's possession (such as an existing health and safety file, an asbestos survey, structural drawings

Guidance 4, 5

etc) or which is reasonable to obtain through sensible enquiry (regulation 2(1)). The information must be relevant to the project, have an appropriate level of detail and be proportionate to the nature of the risks.

43   The client has the main duty for providing pre-construction information. This must be provided as soon as practicable to each designer (including the principal designer) and contractor (including the principal contractor) who is bidding for work on the project or has already been appointed. For projects involving more than one contractor, the client should expect the principal designer to help bring the pre-construction information together and provide it to the designers and contractors involved. Appendix 2 gives further guidance on the requirements relating to pre-construction information. Appendix 5 shows how pre-construction information relates to and influences other types of information during a construction project involving more than one contractor.

The construction phase plan

44   The client must ensure that a construction phase plan for the project is prepared before the construction phase begins. The plan outlines the health and safety arrangements, site rules and specific measures concerning any work involving the particular risks listed in Schedule 3 of CDM 2015. For single-contractor projects, the contractor must ensure the plan is prepared. For projects involving more than one contractor, it is the principal contractor's duty. See Appendix 3 for further guidance on the requirements relating to construction phase plans and Appendix 5 for how a construction phase plan relates to and influences other types of information during a construction project involving more than one contractor.

The health and safety file

45   **A health and safety file is only required for projects involving more than one contractor.** The client must ensure that the principal designer prepares a health and safety file for their project. Its purpose is to ensure that, at the end of the project, the client has information that anyone carrying out subsequent construction work on the building will need to know about in order to be able to plan and carry out the work safely and without risks to health.

46   To ensure that an appropriate health and safety file is produced at the end of the project, the client must:

(a) provide the principal designer with any existing file produced as part of an earlier project so the information it contains can be used to plan the pre-construction phase of the current project;
(b) ensure the principal designer prepares a new file (or revises any existing one);
(c) ensure the principal designer reviews and revises the file regularly and passes the completed file back at the end of the project;
(d) ensure the file is handed to the principal contractor if the principal designer's appointment finishes before the end of the project;
(e) ensure the file is kept available for anyone who needs it to comply with relevant legal requirements; and
(f) pass the file to whoever takes over the building and takes on the client duties if the client decides to dispose of their interest in it.

Appendix 4 gives further guidance on the requirements relating to the health and safety file. Appendix 5 shows how the health and safety file relates to and influences other types of information during a construction project involving more than one contractor.

> **Regulation 6** sets out the duty that a client has to notify the relevant enforcing authority of certain construction projects.

Regulation 6 Notification

Regulation 6

(1) A project is notifiable if the construction work on a construction site is scheduled to—

(a) last longer than 30 working days and have more than 20 workers working simultaneously at any point in the project; or
(b) exceed 500 person days.

(2) Where a project is notifiable, the client must give notice in writing to the Executive as soon as is practicable before the construction phase begins.

(3) The notice must—

(a) contain the particulars specified in Schedule 1;
(b) be clearly displayed in the construction site office in a comprehensible form where it can be read by any worker engaged in the construction work; and
(c) if necessary, be periodically updated.

(4) Where a project includes construction work of a description for which the Office of Rail Regulation is the enforcing authority by virtue of regulation 3 of the Health and Safety (Enforcing Authority for Railways and Other Guided Transport Systems) Regulations 2006, the client must give notice to the Office of Rail Regulation instead of the Executive.

(5) Where a project includes construction work on premises which are or are on—

(a) a GB nuclear site (within the meaning given in section 68 of the Energy Act 2013);
(b) an authorised defence site (within the meaning given in regulation 2(1) of the Health and Safety (Enforcing Authority) Regulations 1998); or
(c) a new nuclear build site (within the meaning given in regulation 2A of those Regulations),

the client must give notice to the Office for Nuclear Regulation instead of the Executive.

Guidance 6

Responsibility for notification

47 Where a construction project must be notified, the client must submit a notice in writing to the relevant enforcing authority (HSE, Office of Rail Regulation (ORR) or Office for Nuclear Regulation (ONR) – see paragraph 50). Every day construction work is likely to take place (including weekends and bank holidays) counts towards the period of construction work.

48 If a construction project is not notifiable at first, but there are subsequent changes to its scope so that it fits the criteria for notification, the client must notify the work to the relevant enforcing authority as soon as possible.

49 The client must submit the notice as soon as practicable before the construction phase begins. In practice, the client may request someone else do this

Guidance 6

on their behalf. Any modifications or updates to the notification should be sent, making clear that they relate to an earlier notification.

50 Details about the information that should be notified are set out in Schedule 1 of CDM 2015. The easiest way to notify any project (to HSE, ORR or ONR) is to use the electronic F10 notification form at www.hse.gov.uk/forms/notification/f10.htm.

51 The client must ensure that an up-to-date copy of the notice is displayed in the construction site office, so it is accessible to anyone working on the site and in a form that can be easily understood. The client can either do this themselves, or ask the principal contractor or contractor to do so.

52 **Remember – the requirements of CDM 2015 apply whether or not the project is notifiable**.

> **Regulation 7** sets out provisions which limit the extent to which domestic clients must carry out the client duties in CDM 2015. Most of the duties are passed to other dutyholders.

Regulation 7 Application to domestic clients

Regulation 7

(1) Where the client is a domestic client the duties in regulations 4(1) to (7) and 6 must be carried out by—

(a) the contractor for a project where there is only one contractor;
(b) the principal contractor for a project where there is more than one contractor; or
(c) the principal designer where there is a written agreement that the principal designer will fulfil those duties.

(2) If a domestic client fails to make the appointments required by regulation 5—

(a) the designer in control of the pre-construction phase of the project is the principal designer;
(b) the contractor in control of the construction phase of the project is the principal contractor.

(3) Regulation 5(3) and (4) does not apply to a domestic client.

Guidance 7

Who is a domestic client?

53 A domestic client is someone who has construction work done on their own home, or the home of a family member, which is **not** done in connection with a business. Local authorities, housing associations, charities, landlords and other businesses may own domestic properties, but they are not a domestic client for the purposes of CDM 2015. If the work is in connection with a business attached to domestic premises, such as a shop, the client is not a domestic client.

| Guidance | 7 | **What should a domestic client do?**

54   A domestic client is not required to carry out the duties placed on commercial clients in regulations 4 (Client duties for managing projects), 6 (Notification) and 8 (General duties) – see also paragraph 23. Where the project involves:

(a) **only one contractor**, the contractor must carry out the client duties as well as the duties they already have as contractor (see paragraphs 147–179). In practice, this should involve doing little more to manage the work to ensure health and safety;

(b) **more than one contractor**, the principal contractor must carry out the client duties as well as the duties they already have as principal contractor (see paragraphs 110–146). If the domestic client has not appointed a principal contractor, the duties of the client must be carried out by the contractor in control of the construction work.

55   In some situations, domestic clients wishing to extend, refurbish or demolish parts of their own property will, in the first instance, engage an architect or other designer to produce possible designs for them. It is also recognised that construction work does not always follow immediately after design work is completed. If they so wish, a domestic client has the flexibility of agreeing (in writing) with their designer that the designer coordinates and manages the project, rather than this role automatically passing to the principal contractor. Where no such agreement is made, the principal contractor will automatically take over the project management responsibilities (see paragraph 54).

56   See Appendix 6 for guidance for dutyholders who work for domestic clients. This also includes a flow chart in Figure 1 showing the transfer of the client duties from a domestic client to other dutyholders involved.

PART 3 Health and safety duties and roles

> **Regulation 8** sets out a number of requirements on anyone working on a project with certain responsibilities. They relate to appointing designers and contractors, the need for cooperation between dutyholders, reporting anything likely to endanger health and safety and ensuring information and instruction is understandable.

Regulation 8 General duties

Regulation 8

(1) A designer (including a principal designer) or contractor (including a principal contractor) appointed to work on a project must have the skills, knowledge and experience and, if they are an organisation, the organisational capability, necessary to fulfil the role that they are appointed to undertake, in a manner that secures the health and safety of any person affected by the project.

(2) A designer or contractor must not accept an appointment to a project unless they fulfil the conditions in paragraph (1).

(3) A person who is responsible for appointing a designer or contractor to carry out work on a project must take reasonable steps to satisfy themselves that the designer or contractor fulfils the conditions in paragraph (1).

(4) A person with a duty or function under these Regulations must cooperate with any other person working on or in relation to a project, at the same or an adjoining construction site, to the extent necessary to enable any person with a duty or function to fulfil that duty or function.

(5) A person working on a project under the control of another must report to that person anything they are aware of in relation to the project which is likely to endanger their own health or safety or that of others.

(6) Any person who is required by these Regulations to provide information or instruction must ensure the information or instruction is comprehensible and provided as soon as is practicable.

(7) To the extent that they are applicable to a domestic client, the duties in paragraphs (3), (4) and (6) must be carried out by the person specified in regulation 7(1).

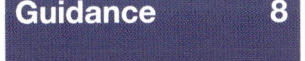

57 Regulation 8 applies to anyone working on a project.

Appointment of designers and contractors

Appointing designers and contractors

58 Anyone appointing a designer or contractor to work on a project must take reasonable steps to satisfy themselves that those who will carry out the work have

Guidance 8

the skills, knowledge, experience, and, where they are an organisation, the organisational capability to carry out the work in a way that secures health and safety. Reasonable steps will depend on the complexity of the project and the range and nature of the risks involved.

59   Organisational capability means the policies and systems an organisation has in place to set acceptable health and safety standards which comply with the law, and the resources and people to ensure the standards are delivered.

60   When appointing a designer or a contractor, sensible and proportionate enquiries should be made about their organisational capability to carry out the work. Only enquiries for information that will address the anticipated risks and capability of the supplier should be made – excessive or duplicated paperwork should be avoided because it can distract attention from the practical management of risks. Those making appointments will find the standard health and safety questions in PAS 91:2013 (Publicly Available Specification) *Construction related procurement. Prequalification questionnaires*[2] a useful aid. Using PAS 91 standard questions is one way of helping to assess organisational capability.

61   As well as carrying out pre-qualification checks on organisations, those responsible for making appointments should also check that the designer or contractor has enough experience and a good record in managing the risks involved in projects. These checks should ideally be carried out at the final stage after pre-qualification checks have been completed and before appointments are made.

62   When considering the requirements for designers and other construction professionals, due weight should also be given to membership of an established professional institution or body. For example, do these bodies have arrangements in place which provide some reassurance that health and safety is part of the route to membership of their profession? However, questions should also be asked of individuals to ensure that they have sufficient skills, knowledge, and experience to carry out the work involved, and that they keep those capabilities up to date.

Designers and contractors seeking appointment

63   Designers and contractors (including individuals and sole traders) must be able to demonstrate they have the health and safety skills, knowledge and experience to carry out the work for which they are seeking appointment. This is the case for individuals working for larger organisations or for themselves – in particular, self-employed designers.

64   Designers or contractors can use the services of an independent (third party) assessor to assess their organisational capability. If they do, there are companies that provide pre-qualification assessment services, including those who are members of the Safety Schemes in Procurement (SSIP) Forum. The SSIP Forum is an umbrella body with binding agreements to ensure member schemes recognise each other's pre-qualification assessments. The website (www.ssip.org.uk) provides a free search facility for any business that has undergone an SSIP assessment and gives further information about SSIP. SSIP assessment is one way a designer or contractor can demonstrate organisational capability at the pre-qualification stage of the appointment process, **but not the only way**.

65   The law does not require any business or individual to use the services of a third party to help them bid for work. Rather than use the services of a third party, it is equally acceptable for a business or individual to assess their own capability and supply relevant documentation to a client in support of a bid for work. The standard health and safety pre-qualification questions in PAS 91 may be helpful in carrying out a self-assessment (see paragraph 60).

Guidance 8

Cooperating with each other

66 Everyone with duties under CDM 2015 must cooperate with others involved with the project or any project on an adjoining site. This means working with each other to ensure health and safety for all concerned. This should involve communicating with others and understanding what they are doing and in what sequence, eg by holding regular coordination and progress meetings.

67 For lower-risk projects involving more than one contractor, a low-key approach will be sufficient. In higher risk projects, a more rigorous approach to cooperation, coordination and planning will be needed. There must also be effective communications between different organisations where they work in close proximity on the same site or on adjoining sites, eg daily updates to make sure there is a common understanding of the work being planned. In all cases, action taken should be in proportion to the risks the construction work activity presents.

Reporting dangerous conditions

68 Everyone involved in a project (including workers) has a duty to report instances where they or others are working in a way that puts them or anyone else in danger. Any instances must be reported to the person in control of the work. The person in control should encourage workers to stop work and report dangerous conditions when they see them.

Providing clear information or instructions

69 Anyone with a duty under CDM 2015 to provide health and safety information or instructions to anyone else must ensure that it is easy to understand. **Information** about hazards is essential to all project workers and managers to make sure they understand the risks involved with the work. **Instructions** are those agreed actions that must be followed to prevent or minimise those risks.

70 Any information or instruction provided should be in simple, clear English (and/or other languages where appropriate). It should also be set out in a logical order and have illustrations where appropriate. The use of photographs or diagrams in written communication can be very helpful. The amount of detail provided should be proportionate to the scale and complexity of the project, the risks and the nature and purpose of the messages. Only information that is necessary to help prevent harm should be provided – unnecessary information can prevent the clear communication of key messages. Examples of types of information include the:

(a) pre-construction information the client is required to provide to designers and contractors;
(b) health and safety information about the design that designers are required to provide to other dutyholders;
(c) information the principal designer must provide to enable preparation of the construction phase plan;
(d) site rules that are part of the construction phase plan; and
(e) information that principal contractors must provide to workers (or workers' representatives).

71 Information or instructions must be provided in good time – before the work begins, so that the recipients can understand and take account of it in carrying out their duties. Wherever possible, it should be made available directly to the people carrying out the work. Where this is not possible, dutyholders and workers need to know what information is available and where it can be found.

Managing health and safety in construction

> **Regulations 9 and 10** set out the duties placed on designers. These include the duty to eliminate, reduce or control foreseeable health and safety risks through the design process, such as those that may arise during construction work or in maintaining and using the building once it is built.

Regulation 9 Duties of designers

Regulation 9

(1) A designer must not commence work in relation to a project unless satisfied that the client is aware of the duties owed by the client under these Regulations.

(2) When preparing or modifying a design the designer must take into account the general principles of prevention and any pre-construction information to eliminate, so far as is reasonably practicable, foreseeable risks to the health or safety of any person—

(a) carrying out or liable to be affected by construction work;
(b) maintaining or cleaning a structure; or
(c) using a structure designed as a workplace.

(3) If it is not possible to eliminate these risks, the designer must, so far as is reasonably practicable—

(a) take steps to reduce or, if that is not possible, control the risks through the subsequent design process;
(b) provide information about those risks to the principal designer; and
(c) ensure appropriate information is included in the health and safety file.

(4) A designer must take all reasonable steps to provide, with the design, sufficient information about the design, construction or maintenance of the structure, to adequately assist the client, other designers and contractors to comply with their duties under these Regulations.

Regulation 10 Designs prepared or modified outside Great Britain

Regulation 10

(1) Where a design is prepared or modified outside Great Britain for use in construction work to which these Regulations apply—

(a) the person who commissions it, if established within Great Britain; or
(b) if that person is not so established, the client for the project,

must ensure that regulation 9 is complied with.

(2) This regulation does not apply to a domestic client.

Guidance 9, 10

Who is a designer?

72 A designer is an organisation or individual, who:

(a) prepares or modifies a design for a construction project (including the design of temporary works); or
(b) arranges for, or instructs someone else to do so.

Guidance 9, 10

The term 'design' includes drawings, design details, specifications, bills of quantity and calculations prepared for the purpose of a design. Designers include architects, architectural technologists, consulting engineers, quantity surveyors, interior designers, temporary work engineers, chartered surveyors, technicians or anyone who specifies or alters a design. They can include others if they carry out design work, such as principal contractors, and specialist contractors, eg an engineering contractor providing design, procurement and construction management services. Where commercial clients become actively involved in designing in relation to their project, they may also be considered to be designers.

73 Local authority or government officials may give advice and instruction on designs meeting statutory requirements (eg the Building Regulations), but this does not make them designers. A designer may have no choice but to comply with these requirements, which are a 'design constraint'. However, if statutory bodies ask for particular features to be included or excluded which go beyond what the law requires (eg stipulating the absence of edge protection on flat roofs if there is no basis in planning law or policies to do so), they may become designers under CDM 2015 and must comply with its requirements.

74 The person who selects products for use in construction is a designer and must take account of health and safety issues arising from their use. If a product is purpose-built, the person who prepares the specification is a designer and so are manufacturers, if they develop a detailed design.

Why is a designer important?

75 A designer has a strong influence during the concept and feasibility stage of a project. The earliest decisions can fundamentally affect the health and safety of those who will construct, maintain, repair, clean, refurbish and eventually demolish a building. The health and safety of those who use a building as a workplace may also be affected. Decisions such as selecting materials that are lighter to handle or windows that can be cleaned from the inside can avoid or reduce the risks involved in constructing the building and maintaining it after construction. Although it is understood that residual risks may well remain, decisions such as these have an important influence on the overall health and safety performance of the project and the use and maintenance of the building once it is built.

76 **A designer should address health and safety issues from the very start.** Where issues are not addressed early on, projects can be delayed and it can become significantly harder for contractors to devise safe ways of working once they are on site. The client may also be forced to make costly late changes, so the building can be used and maintained safely once it is built.

When do a designer's duties apply?

77 The designer's duties apply as soon as designs which may be used in construction work in Great Britain are started. This includes concept design, competitions, bids for grants, modification of existing designs and relevant work carried out as part of feasibility studies. It does not matter whether planning permission or funds have been secured, or the client is a domestic client.

78 If a design is prepared or modified outside Great Britain, the designer duties apply to the person or organisation who commissions it if they are established in Great Britain, or if not, the client (but not a domestic client).

Guidance 9, 10

What must a designer do?

Making clients aware of their duties

79 A designer must not start any design work unless they are satisfied the client is aware of the duties clients have under CDM 2015. This duty can be fulfilled as part of routine business, eg in early meetings or liaison with the client to discuss the project. A designer should have a sufficient knowledge of client duties to give sufficient advice about the project. The level of advice will depend on the knowledge and experience of the client and the complexities of the project.

80 On projects involving more than one contractor, the task of informing the client of their duties should normally fall to the principal designer. Any other designers appointed can seek confirmation from the principal designer that the client has been made aware of their duties.

Preparing or modifying designs

81 When preparing or modifying designs, a designer must take account of the general principles of prevention, and the pre-construction information provided to them, with the aim, as far as reasonably practicable, of **eliminating** foreseeable risks. Where this is not possible they must take reasonably practicable steps to **reduce** the risks or **control** them through the design process, and provide information about the remaining risks to other dutyholders. See paragraphs 82–90 for further guidance.

Taking account of the general principles of prevention in design work

82 The general principles of prevention are set out in Appendix 1 and provide a framework within which designers must consider their designs and any potential risks which may affect:

(a) workers or anyone else (eg members of the public) who may be affected during construction;
(b) those who may maintain or clean the building once it is built; or
(c) those who use the building as a workplace.

Designs prepared for places of work also need to comply with the Workplace (Health, Safety and Welfare) Regulations 1992 (the Workplace Regulations),[3] taking account of factors such as lighting and the layout of traffic routes.

83 Health and safety risks need to be considered alongside other factors that influence the design, such as cost, fitness for purpose, aesthetics and environmental impact. Working with contractors (including principal contractors) involved in the project can help identify the potential risks and ways they may be controlled.

84 Once the risks have been considered, the level of detail in the information provided to those who need it should be proportionate to the risks remaining. Insignificant risks can usually be ignored, as can those arising from routine construction activities, unless the design worsens or significantly alters these risks.

Taking account of pre-construction information

85 A designer must take account of pre-construction information the client or principal designer provides when making decisions about the extent to which they can eliminate foreseeable risks through the designs they produce; and, where these risks cannot be eliminated, the steps they take to reduce or control them. Appendix 2 gives further guidance on the requirements relating to pre-construction information. Appendix 5 shows how pre-construction information relates to and influences other types of information during a construction project involving more than one contractor.

| Guidance 9, 10 | *Eliminating, reducing or controlling foreseeable risks through design* |

86 When designing, a designer must consider the risks people may be exposed to through the course of both constructing a building and using it once it is constructed. Designing is a process that often continues throughout the project and the following questions should be considered when design is carried out:

(a) Can I get rid of the problem (or hazard) altogether? For example, can air-conditioning plant on a roof be moved to ground level, so work at height is not required for either installation or maintenance?

(b) If not, how can I reduce or control the risks, so that harm is unlikely or the potential consequences less serious? For example, can I place the plant within a building on the roof, or provide a barrier around the roof?

87 If risks cannot be eliminated altogether, a designer should apply the principles below in deciding how to reduce or control the remaining risks – if possible, in the following order:

(a) provide a less risky option, eg switch to using paving lighter in weight, to reduce musculoskeletal disorders such as back problems;

(b) make provisions so the work can be organised to reduce exposure to hazards, eg make provision for traffic routes so barriers can be provided between pedestrians and traffic;

(c) ensure that those responsible for planning and managing the work are given the information they will need to manage remaining risks, eg tell them about loads that will be particularly heavy or elements of the building that could become unstable. This can be achieved through providing key information on drawings or within models, eg by using Building Information Modelling (BIM).

88 When addressing risks, a designer is expected to do as much as is reasonable at the time the design is prepared. Risks that cannot be addressed at the initial stage of a project may need to be reviewed later on during detailed design. On projects involving more than one contractor, the principal designer will lead in managing the review process. Further information on eliminating, reducing or controlling foreseeable risks is available on HSE's construction web pages www.hse.gov.uk/construction/areyou/designer.

Providing design information

89 A designer must provide information to other dutyholders using or implementing the design. This includes information for:

(a) the principal designer:
  (i) about significant risks (see Glossary) associated with the design that cannot be eliminated, so it can form part of the pre-construction information (see Appendices 2 and 5 for further guidance);
  (ii) to take into account in preparing or revising the health and safety file (see Appendix 4 for further guidance);
(b) other designers;
(c) the principal contractor (or the contractor on a single-contractor project) who has responsibility for preparing, reviewing and revising the construction phase plan for the project (see Appendix 3 for further guidance); and
(d) contractors who construct the design.

90 The designer should agree with the principal designer the arrangements for sharing information to avoid omissions or duplicated effort. Those who need the information should be given it at the right time. For example, in preparing the construction phase plan, the information should be provided well before the construction phase begins.

Guidance   **9, 10**

Cooperating with other dutyholders

91   Designers should liaise with any other designers, including the principal designer, so that work can be coordinated to establish how different aspects of designs interact and influence health and safety. This includes temporary and permanent works designers. Designers must also cooperate with contractors and principal contractors so that their knowledge and experience about, eg the practicalities of building the design, is taken into account.

92   Depending on the nature and extent of design work, there may be a need to carry out design reviews. Reviews enable the project team to focus on health and safety matters alongside other key aspects of the project. This can be done as part of the normal design process. The need for such reviews is likely to continue throughout the project although their frequency and the level of detail covered should remain proportionate to the scale and complexity of the design work.

Working for domestic clients

93   A designer's role on a project for a domestic client is no different to the role undertaken for commercial clients. The designer must still carry out their duties to the extent necessary given the risks involved in the project. However, regulation 7 transfers the duties of the domestic client to another dutyholder (which dutyholder depends on the nature of the project) and designers will work to that dutyholder as 'client' for the project. Further guidance is at Appendix 6.

> **Regulation 11** sets out the duties a principal designer has during the pre-construction phase. They include requirements to plan, manage, monitor and coordinate health and safety during this phase and to liaise with the principal contractor in providing information relevant for the planning, management and monitoring of the construction phase.

Regulation 11 Duties of a principal designer in relation to health and safety at the pre-construction phase

Regulation   **11**

(1)   The principal designer must plan, manage and monitor the pre-construction phase and coordinate matters relating to health and safety during the pre-construction phase to ensure that, so far as is reasonably practicable, the project is carried out without risks to health or safety.

(2)   In fulfilling the duties in paragraph (1), and in particular when—

(a)   design, technical and organisational aspects are being decided in order to plan the various items or stages of work which are to take place simultaneously or in succession; and
(b)   estimating the period of time required to complete such work or work stages,

the principal designer must take into account the general principles of prevention and, where relevant, the content of any construction phase plan and health and safety file.

(3)   In fulfilling the duties in paragraph (1), the principal designer must identify and eliminate or control, so far as is reasonably practicable, foreseeable risks to the health or safety of any person—

Regulation 11

(a) carrying out or liable to be affected by construction work;
(b) maintaining or cleaning a structure; or
(c) using a structure designed as a workplace.

(4) In fulfilling the duties in paragraph (1), the principal designer must ensure all designers comply with their duties in regulation 9.

(5) In fulfilling the duty to coordinate health and safety matters in paragraph (1), the principal designer must ensure that all persons working in relation to the pre-construction phase cooperate with the client, the principal designer and each other.

(6) The principal designer must—

(a) assist the client in the provision of the pre-construction information required by regulation 4(4); and
(b) so far as it is within the principal designer's control, provide pre-construction information, promptly and in a convenient form, to every designer and contractor appointed, or being considered for appointment, to the project.

(7) The principal designer must liaise with the principal contractor for the duration of the principal designer's appointment and share with the principal contractor information relevant to the planning, management and monitoring of the construction phase and the coordination of health and safety matters during the construction phase.

Guidance 11

Who is a principal designer?

94 A principal designer is the designer as defined in regulation 2(1) (see also paragraphs 72–74) with control over the pre-construction phase of the project. This is the very earliest stage of a project from concept design through to planning the delivery of the construction work. The principal designer must be appointed in writing by the client.

95 The principal designer can be an organisation or an individual that has:

(a) the technical knowledge of the construction industry relevant to the project;
(b) the skills, knowledge and experience to understand, manage and coordinate the pre-construction phase, including any design work carried out after construction begins.

Where the principal designer is an organisation, it must have the organisational capability to carry out the role.

96 Principal designers may have separate duties as designers (see paragraphs 79–93).

Why is the principal designer important?

97 In liaison with the client and principal contractor, the principal designer has an important role in influencing how the risks to health and safety should be managed and incorporated into the wider management of a project. Decisions about the design taken during the pre-construction phase can have a significant effect on whether the project is delivered in a way that secures health and safety. The principal designer's role involves coordinating the work of others in the project team

Guidance 11

to ensure that significant and foreseeable risks are managed throughout the design process.

What must a principal designer do?

Planning, managing, monitoring and coordinating the pre-construction phase

98 In carrying out the duty to plan, manage, monitor and coordinate the pre-construction phase, principal designers must take account of the general principles of prevention (see paragraph 5 and Appendix 1) and, where relevant, the content of:

(a) pre-construction information (see Appendix 2);
(b) any construction phase plan (see Appendix 3). This will be relevant when the plan has implications for design work carried out after the construction phase has started, eg ground contamination discovered affecting the choice of piling method; and
(c) any existing health and safety file (see Appendix 4). In cases where a health and safety file has been prepared as part of previous construction work on the building, it should have information which will help the planning, management and coordination of the pre-construction phase.

This information should be taken into account particularly when decisions are being taken about design, technical and organisational issues to plan which items or stages of work can take place at the same time or in what sequence; and when estimating the time needed to complete certain items or stages of work.

99 The principal designer's work should focus on ensuring the design work in the pre-construction phase contributes to the delivery of positive health and safety outcomes. Bringing together designers as early as possible in the project, and then on a regular basis, to ensure everyone carries out their duties, will help to achieve this. This can be done as part of the normal design process. Regular design meetings chaired by the principal designer are an effective way to:

(a) discuss the risks that should be addressed during the pre-construction phase;
(b) decide on the control measures to be adopted; and
(c) agree the information that will help prepare the construction phase plan.

100 If the principal designer appoints any designers they must check they have sufficient skills, knowledge, experience and (if they are an organisation) the organisational capability to carry out the work. These checks should be carried out before appointment (see paragraphs 58–62 for further guidance).

101 The principal designer's role continues into the construction phase when design work is carried out and when gathering and preparing information for the health and safety file.

Identifying, eliminating or controlling foreseeable risks

102 Principal designers must ensure, as far as reasonably practicable, that foreseeable risks to health and safety are identified. In practice, this will involve the principal designer working with other designers involved with the project. The risks that should be identified are the significant ones and which are likely to arise:

(a) while carrying out construction work; or
(b) during maintenance, cleaning or using the building as a workplace once it is built.

Guidance 11

Identifying insignificant risks is not an effective way of alerting other dutyholders to the important design issues they need to know about. Designers should be able to demonstrate they have addressed only the significant risks.

103 Once the risks have been identified, principal designers must follow the approach to managing them set out in the general principles of prevention (see Appendix 1). The principal designer must, as far as reasonably practicable, ensure that the design team:

(a) **eliminate** the risks associated with design elements.

If this is not possible (for instance because of competing design considerations such as planning restrictions, specifications, disproportionate costs or aesthetics):

(b) **reduce** any remaining risks; or
(c) **control** them,

to an acceptable level. This relies on exercising judgement in considering how to manage the risks. The focus should be on those design elements where there is a significant risk of injury or ill health.

Ensuring coordination and cooperation

104 Principal designers must ensure as far as reasonably practicable that:

(a) everyone involved in working on the pre-construction phase cooperates with each other. They must establish that effective communication is occurring and that information is shared within the project team. This could involve holding meetings with others in the design team. Progress meetings with the client and the principal contractor also provide a way of ensuring work on the project is properly coordinated;
(b) designers comply with their duties. Appropriate checks should be made to ensure designers are dealing with design risks appropriately. This can be done as part of the design process and through regular progress meetings;
(c) designers provide information about elements of the design which present significant risks that cannot be eliminated. This should include information about unusual or complex risks that are more likely to be missed or misunderstood by contractors or others on the project rather than risks that are well known and understood.

Providing pre-construction information

105 Pre-construction information is information already in the client's possession or which is reasonably obtainable. It must be relevant, have an appropriate level of detail and be proportionate to the nature of risks involved in the project.

106 The client has responsibility for pre-construction information (see paragraphs 42–43). The principal designer must help the client bring together the information the client already holds (such as any existing health and safety file or asbestos survey). The principal designer should then:

(a) assess the adequacy of existing information to identify any gaps in the information which it is necessary to fill;
(b) provide advice to the client on how the gaps can be filled and help them in gathering the necessary additional information; and
(c) provide, as far as they are able to, the additional information promptly and in a convenient form to help designers and contractors who:
　(i) are being considered for appointment; or
　(ii) have already been appointed, to carry out their duties.

Managing health and safety in construction

Guidance 11

Appendix 2 gives further guidance on the requirements relating to pre-construction information. Appendix 5 shows how pre-construction information relates to and influences other types of information during a construction project involving more than one contractor.

Liaising with the principal contractor

107 The principal designer must liaise with the principal contractor for the duration of their appointment. During the pre-construction phase this must cover sharing information that may affect the planning, management, monitoring and coordination of the construction phase – in particular, the information needed by the principal contractor to prepare the construction phase plan (see Appendix 3). Liaison should also extend into the construction phase to deal with ongoing design and obtaining information for the health and safety file. This could be done by holding regular progress meetings with the principal contractor.

108 If the principal designer's appointment finishes before the end of the project, they must ensure that the principal contractor has all the relevant information so that the principal contractor:

(a) is aware of the risks which have not been eliminated in the designs;
(b) understands the means employed to reduce or control those risks; and
(c) understands the implications for implementing the design work for the rest of the project.

The principal designer should also arrange a handover of the health and safety file to the principal contractor and make them aware of any issues to take into account when reviewing, updating and completing it.

Working for domestic clients

109 A principal designer's role when working on a project for a domestic client is no different to their role when carrying out work for a commercial client. They must still carry out the duties set out in regulations 8, 11 and 12 in proportion to the risks involved in the project. But, the effect of regulation 7 is to transfer the duties of the domestic client to another dutyholder. This can be the principal designer when the domestic client chooses to enter into a written agreement with the principal designer to transfer the client duties to them. See Appendix 6 for guidance on how this affects what principal designers must do on domestic projects.

> **Regulation 12** sets out the duties on either the principal designer or principal contractor for the preparation, review, revision and updating of construction phase plans and health and safety files.

Regulation 12 Construction phase plan and health and safety file

Regulation 12

(1) During the pre-construction phase, and before setting up a construction site, the principal contractor must draw up a construction phase plan or make arrangements for a construction phase plan to be drawn up.

(2) The construction phase plan must set out the health and safety arrangements and site rules taking account, where necessary, of the industrial activities taking place on the construction site and, where applicable, must include specific measures concerning work which falls within one or more of the categories set out in Schedule 3.

Regulation 12

(3) The principal designer must assist the principal contractor in preparing the construction phase plan by providing to the principal contractor all information the principal designer holds that is relevant to the construction phase plan including—

(a) pre-construction information obtained from the client;
(b) any information obtained from designers under regulation 9(3)(b).

(4) Throughout the project the principal contractor must ensure that the construction phase plan is appropriately reviewed, updated and revised from time to time so that it continues to be sufficient to ensure that construction work is carried out, so far as is reasonably practicable, without risks to health or safety.

(5) During the pre-construction phase, the principal designer must prepare a health and safety file appropriate to the characteristics of the project which must contain information relating to the project which is likely to be needed during any subsequent project to ensure the health and safety of any person.

(6) The principal designer must ensure that the health and safety file is appropriately reviewed, updated and revised from time to time to take account of the work and any changes that have occurred.

(7) During the project, the principal contractor must provide the principal designer with any information in the principal contractor's possession relevant to the health and safety file, for inclusion in the health and safety file.

(8) If the principal designer's appointment concludes before the end of the project, the principal designer must pass the health and safety file to the principal contractor.

(9) Where the health and safety file is passed to the principal contractor under paragraph (8), the principal contractor must ensure that the health and safety file is appropriately reviewed, updated and revised from time to time to take account of the work and any changes that have occurred.

(10) At the end of the project, the principal designer, or where there is no principal designer the principal contractor, must pass the health and safety file to the client.

Guidance 12

110 Paragraphs 111–115 summarise the actions principal contractors and principal designers must take to ensure the preparation of the construction phase plan and health and safety file and to ensure they remain fit for purpose. See Appendices 3 and 4 for detailed guidance on the responsibilities of all dutyholders in relation to the plan and the file. See Appendix 5 for how the plan and the file relate to and influence other types of information during a construction project involving more than one contractor.

The construction phase plan

111 The construction phase plan must set out the arrangements for securing health and safety during the period construction work is carried out. These arrangements include site rules and any specific measures put in place where work involves one or more of the risks listed in Schedule 3.

112 For projects involving more than one contractor, the principal contractor must ensure the plan is drawn up during the pre-construction phase and **before** the

> **Guidance 12**

construction site is set up. It must take into account the information the principal designer holds, such as the pre-construction information (see Appendix 2) and any information obtained from designers. During the construction phase, the principal contractor must ensure the plan is appropriately reviewed, updated and revised, so it remains effective.

113 For single contractor projects, the contractor must ensure the construction phase plan is drawn up. Guidance on this can be found in paragraph 161 and Appendix 3.

The health and safety file

114 **The health and safety file is only required for projects involving more than one contractor**. It must contain relevant information about the project which should be taken into account when any construction work is carried out on the building **after** the current project has finished. Information included should only be that which is needed to plan and carry out future work safely and without risks to health.

115 The principal designer must prepare the file, and review, update and revise it as the project progresses. If their appointment continues to the end of the project they must also pass the completed file to the client to keep. If the principal designer's appointment finishes before the end of the project, the file must be passed to the principal contractor for the remainder of the project. The principal contractor must then take responsibility for reviewing, updating and revising it and passing it to the client when the project finishes.

> **Regulation 13** sets out the principal contractor's duties during the construction phase. The main duty is to plan, manage, monitor and coordinate health and safety during this phase. Other duties include making sure suitable site inductions and welfare facilities are provided.

Regulation 13 Duties of a principal contractor in relation to health and safety at the construction phase

> **Regulation 13**

(1) The principal contractor must plan, manage and monitor the construction phase and coordinate matters relating to health and safety during the construction phase to ensure that, so far as is reasonably practicable, construction work is carried out without risks to health or safety;

(2) In fulfilling the duties in paragraph (1), and in particular when—

(a) design, technical and organisational aspects are being decided in order to plan the various items or stages of work which are to take place simultaneously or in succession; and

(b) estimating the period of time required to complete the work or work stages;

the principal contractor must take into account the general principles of prevention.

(3) The principal contractor must—

(a) organise cooperation between contractors (including successive contractors on the same construction site);

Regulation 13

(b) coordinate implementation by the contractors of applicable legal requirements for health and safety; and

(c) ensure that employers and, if necessary for the protection of workers, self-employed persons—

(i) apply the general principles of prevention in a consistent manner, and in particular when complying with the provisions of Part 4;

(ii) where required, follow the construction phase plan.

(4) The principal contractor must ensure that—

(a) a suitable site induction is provided;

(b) the necessary steps are taken to prevent access by unauthorised persons to the construction site; and

(c) facilities that comply with the requirements of Schedule 2 are provided throughout the construction phase.

(5) The principal contractor must liaise with the principal designer for the duration of the principal designer's appointment and share with the principal designer information relevant to the planning, management and monitoring of the pre-construction phase and the coordination of health and safety matters during the pre-construction phase.

Guidance 13

Who is a principal contractor?

116 A principal contractor is the organisation or person that coordinates the work of the construction phase of a project involving more than one contractor, so it is carried out in a way that secures health and safety. They are appointed by the client and must possess the skills, knowledge, and experience, and (if an organisation) the organisational capability to carry out their role effectively given the scale and complexity of the project and the nature of the health and safety risks involved.

117 There may be occasions where two or more projects are taking place on the same site at the same time, but are run independently of one another. Whatever the circumstances, **it is essential that there is clarity over who is in control during the construction phase in any part of the site at any given time**. Where it is not possible for one principal contractor to be in overall control, those principal contractors involved must:

(a) cooperate with one another;
(b) coordinate their work; and
(c) take account of any shared interfaces between the activities of each project (eg shared traffic routes).

Why is a principal contractor important?

118 Good management of health and safety on site is crucial to the successful delivery of a construction project. In liaison with the client and principal designer, principal contractors have an important role in managing the risks of the construction work and providing strong leadership to ensure standards are understood and followed.

| **Guidance** | **13** |

What must a principal contractor do?

Planning, managing, monitoring and coordinating the construction phase
General

119 In planning, managing, monitoring and coordinating the construction phase, a principal contractor must take account of the general principles of prevention (see Appendix 1). They must take account of these principles when:

(a) decisions are being taken to plan which items or stages of work can take place at the same time or in sequence; and
(b) estimating the time certain items or stages of work will take to complete.

120 The principal contractor should be appointed by the client before the construction phase begins to allow them to work closely with:

(a) the client for the life of the project; and
(b) the principal designer for the remainder of their appointment.

This work must include liaising with the principal designer for the purposes of planning, managing, monitoring and coordinating the pre-construction phase. As the project moves into the construction phase, the principal contractor should take the lead in planning, managing, monitoring and coordinating the project while continuing to liaise with the client and principal designer.

121 The effort the principal contractor devotes to carrying out their duties should be in proportion to the size and complexity of the project and the risks involved. The principal contractor should expect and receive help from other dutyholders in identifying the risks associated with the work and determining the controls that need to be put in place. In particular:

(a) the client must provide (with help from the principal designer) the pre-construction information (see Appendix 2); and
(b) the principal designer must provide any other information needed for the preparation of the construction phase plan (see Appendix 3).

122 A principal contractor must also ensure anyone they appoint has the skills, knowledge, and experience and, where they are an organisation, the organisational capability to carry out the work in a way that secures health and safety (see paragraphs 58–62).

Planning

123 Planning must take into account the risks to all those affected – workers, members of the public and the client's employees, if working in an occupied premises. It must cover:

(a) the risks likely to arise during construction work;
(b) the measures needed to protect those affected by planning to provide:
 (i) and maintain the right plant and equipment;
 (ii) the necessary information, instruction and training; and
 (iii) the right level of supervision;
(c) the resources (including time) needed to organise and deliver the work, including its management, monitoring and coordination.

124 The pre-construction information (see Appendix 2) and any key design information, identifying risks that need to be managed during construction work, will be helpful in planning the construction phase and drawing up the construction phase plan (see Appendix 3). Appendix 5 shows how pre-construction information

Guidance 13

and the construction phase plan relate to and influence other types of information during a construction project involving more than one contractor.

Managing

125 To manage the construction phase, principal contractors must ensure that:

(a) those engaged to carry out the work are capable of doing so;
(b) effective, preventative and protective measures are put in place to control the risks; and
(c) the right plant, equipment and tools are provided to carry out the work involved.

126 Managing people to prevent and control risk requires leadership. Principal contractors can demonstrate visible leadership through the actions of their managers. These actions include setting standards for working practices and providing an example by following them. Leaders in health and safety should have a strong grasp of what is needed in a given situation, make clear decisions, and be able to communicate effectively.

127 A systematic approach to managing should be taken to ensure workers understand:

(a) the risks and control measures on the project;
(b) who has responsibility for health and safety;
(c) that consistent standards apply throughout the project and will be checked frequently;
(d) where they can locate health and safety information which is easily understandable, well organised and relevant to the site; and
(e) that incidents will be investigated and lessons learned.

128 Good supervision is part of showing leadership in health and safety. It:

(a) focuses workers' attention on risks, and how to prevent them;
(b) shows commitment to establishing and maintaining the control measures;
(c) involves consulting effectively with workers, taking into account their views; and
(d) challenges unsafe conditions and working practices when they arise.

Principal contractors do not have to undertake detailed supervision of contractors' work.

Monitoring

129 Standards should be checked regularly given the rapidly changing nature of a construction site. Effective monitoring involves:

(a) time and effort (with sufficient resource having been set aside for this at the planning stage – see paragraph 123);
(b) treating health and safety in the same way as other important aspects of the business;
(c) taking prompt action where necessary; and
(d) using a mix of performance measures – both active and reactive in nature, eg:
  (i) routine checks of site access and work areas and plant and equipment, or health risk management to prevent harm **(active)**;
  (ii) investigating near-miss incidents and injuries as well as monitoring cases of ill health **(reactive)**.

Guidance 13

Coordinating

130 A principal contractor has a specific duty to ensure that contractors under their control cooperate with each other so the risks to themselves and others affected by the work are managed effectively. This includes ensuring contractors who start work at different stages of the construction phase cooperate with each other so any information and instruction relevant for a new contractor to carry out their work safely is provided to them. Regular planning meetings between the principal contractor and contractors are an effective way of ensuring this.

131 The need for coordination does not just apply when implementing the requirements in CDM 2015 but also when complying with any other health and safety requirements. In coordinating the work of employers and self-employed under their control, principal contractors must ensure they:

(a) apply the general principles of prevention (see Appendix 1); and
(b) where required, follow the construction phase plan (see Appendix 3).

This will involve the principal contractor liaising with those involved to establish a common understanding of the health and safety standards expected and gaining their cooperation in meeting these standards. The extent to which the principal contractor should liaise will depend on the risks involved.

132 The principal contractor should also work with the client to ensure there is cooperation with others outside the construction site who may be affected by the activities on site. This includes coordinating the activities of contractors on the principal contractor's site with contractors on any neighbouring sites, particularly where the activities on each site combine to create hazards outside the sites that need to be addressed jointly.

Providing suitable site inductions

133 The principal contractor must ensure every site worker is given a suitable site induction. The induction should be site specific and highlight any particular risks (including those listed in Schedule 3) and control measures that those working on the project need to know about. The following issues should be considered:

(a) senior management commitment to health and safety;
(b) outline of the project;
(c) management of the project;
(d) first-aid arrangements;
(e) accident and incident reporting arrangements;
(f) arrangements for briefing workers on an ongoing basis, eg toolbox talks;
(g) arrangements for consulting the workforce on health and safety matters;
(h) individual worker's responsibility for health and safety.

134 Site inductions should also be provided to those who do not regularly work on the site, but who visit it on an occasional (eg architects) or once-only basis (eg students). **The inductions should be proportionate to the nature of the visit.** Inductions provided to escorted visitors need not have the detail that unescorted visitors should have. Escorted visitors only need to be made aware of the main hazards they may be exposed to and the control measures.

Preventing unauthorised access to the site

135 The principal contractor must ensure reasonable steps are taken to prevent unauthorised access onto the construction site. They should liaise with the contractors on site to physically define the site boundaries by using suitable barriers which take account of the nature of the site and its surrounding environment. The principal contractor should also take steps to ensure that only those authorised to access the site do so.

Guidance 13

136 Special consideration will be required for sites that have:

(a) rights of way through them;
(b) other work areas next to them, eg a shop refurbishment in a shopping centre;
(c) occupied houses next to them, especially on new-build housing estates;
(d) children or vulnerable people nearby, eg schools or care homes located near the site.

Providing welfare facilities

137 The principal contractor must ensure that suitable and sufficient welfare facilities are provided and maintained throughout the construction phase. What is suitable and sufficient will depend on the size and nature of the workforce involved in the project. Facilities must be made available before any construction work starts and should be maintained until the end of the project. See paragraphs 176–178 and HSE's construction web pages at www.hse.gov.uk/construction for more guidance on welfare facilities.

138 The principal contractor should liaise with other contractors involved with the project to ensure appropriate welfare facilities are provided. Such liaison should continue for the duration of the construction phase and take account of any changes in the nature of the site which require, in turn, changes to the provision of welfare facilities.

Liaising with the principal designer

139 The principal contractor must liaise with the principal designer for the duration of the project. The early appointment of a principal contractor by the client will allow their construction expertise to be used from the earliest stages of designing and planning a project. They should also liaise with the principal designer throughout the construction phase on matters such as changes to the designs and the implications these changes may have for managing the health and safety risks.

140 Liaison should cover drawing together information the principal designer will need:

(a) to prepare the health and safety file (see Appendix 4); or
(b) that may affect the planning and management of the pre-construction phase. The pre-construction information is important for planning and managing this phase and the subsequent development of the construction phase plan (see Appendix 3). Appendix 2 contains further guidance on pre-construction information, including the information principal contractors should consider providing to both the principal designer and the client.

Other duties

141 The principal contractor also has additional duties they must comply with. These include the duty to consult and engage with workers (see paragraphs 143–146).

Working for domestic clients

142 A principal contractor's role when working on a project for a domestic client is no different to their role when carrying out work for a commercial client. They must still carry out the duties set out in regulations 8, 12, 13 and 14 in proportion to the risks involved in the project. But, the effect of regulation 7 is to transfer in certain circumstances (see paragraphs 53–56) the duties of the domestic client to another dutyholder, including the principal contractor. See Appendix 6 for guidance on how this affects what principal contractors must do on domestic projects.

Regulation 14 places duties on the principal contractor to consult and engage with workers or their representatives. These duties are in addition to those in separate legislation which requires employers to consult with their employees (or their representatives) – see paragraph 13.

Regulation 14 Principal contractor's duties to consult and engage with workers

Regulation 14

The principal contractor must—

(a) make and maintain arrangements which will enable the principal contractor and workers engaged in construction work to cooperate effectively in developing, promoting and checking the effectiveness of measures to ensure the health, safety and welfare of the workers;

(b) consult those workers or their representatives in good time on matters connected with the project which may affect their health, safety or welfare, in so far as they or their representatives have not been similarly consulted by their employer;

(c) ensure that those workers or their representatives can inspect and take copies of any information which the principal contractor has, or which these Regulations require to be provided to the principal contractor, which relate to the health, safety or welfare of workers at the site, except any information—

(i) the disclosure of which would be against the interests of national security;

(ii) which the principal contractor could not disclose without contravening a prohibition imposed by or under an enactment;

(iii) relating specifically to an individual, unless that individual has consented to its being disclosed;

(iv) the disclosure of which would, for reasons other than its effect on health, safety or welfare at work, cause substantial injury to the principal contractor's undertaking or, where the information was supplied to the principal contractor by another person, to the undertaking of that other person;

(v) obtained by the principal contractor for the purpose of bringing, prosecuting or defending any legal proceedings.

Guidance 14

143 The importance of involving workers in decisions about health and safety is a vital element to securing health and safety in the construction industry. A principal contractor has a duty under CDM 2015 to involve the workforce in matters of health, safety and welfare. This is in addition to the duty on all employers to consult with their employees (or their representatives) on health and safety matters under separate legislation (see paragraphs 12–15). Further guidance on involving workers can be found in the free *Leadership and Worker Involvement Toolkit* (see paragraph 15).

144 The principal contractor must consult and engage with the workforce to ensure that measures for their health, safety and welfare are developed, promoted and checked for effectiveness. Consultation must be carried out in a timely manner. If consultation has already taken place through a direct employer, it is not required again.

145 Effective worker involvement will develop from effective consultation and cooperation between the principal contractor and other contractors on site. The following techniques help in achieving this:

Guidance 14

(a) commitment by managers to lead by example, to provide the resources and set the standards of health and safety expected;

(b) implementation of a range of ways to communicate, ensure cooperation with and consult the workforce in managing health and safety; and

(c) collecting evidence that worker involvement is effective and that cooperation between contractors is effective.

146 The construction workforce should also have access to, and be able to take copies of, any information the principal contractor has which may affect their health, safety and welfare. The exceptions to this are set out in regulation 14(c).

> **Regulation 15** places duties on contractors. The main duty is to plan, manage and monitor the work under their control. Other duties include:
>
> (a) complying with directions given to them by either the principal designer or principal contractor on sites where there is more than one contractor; and
>
> (b) preparing a construction phase plan on sites where they are the only contractor.

Regulation 15 Duties of contractors

Regulation 15

(1) A contractor must not carry out construction work in relation to a project unless satisfied that the client is aware of the duties owed by the client under these Regulations.

(2) A contractor must plan, manage and monitor construction work carried out either by the contractor or by workers under the contractor's control, to ensure that, so far as is reasonably practicable, it is carried out without risks to health and safety.

(3) Where there is more than one contractor working on a project, a contractor must comply with—

(a) any directions given by the principal designer or the principal contractor; and

(b) the parts of the construction phase plan that are relevant to that contractor's work on the project.

(4) If there is only one contractor working on the project, the contractor must take account of the general principles of prevention when—

(a) design, technical and organisational aspects are being decided in order to plan the various items or stages of work which are to take place simultaneously or in succession; and

(b) estimating the period of time required to complete the work or work stages.

(5) If there is only one contractor working on the project, the contractor must draw up a construction phase plan, or make arrangements for a construction phase plan to be drawn up, as soon as is practicable prior to setting up a construction site.

(6) The construction phase plan must fulfil the requirements of regulation 12(2).

Regulation 15

(7) A contractor must not employ or appoint a person to work on a construction site unless that person has, or is in the process of obtaining, the necessary skills, knowledge, training and experience to carry out the tasks allocated to that person in a manner that secures the health and safety of any person working on the construction site.

(8) A contractor must provide each worker under their control with appropriate supervision, instructions and information so that construction work can be carried out, so far as is reasonably practicable, without risks to health and safety.

(9) The information provided must include—

(a) a suitable site induction, where not already provided by the principal contractor;
(b) the procedures to be followed in the event of serious and imminent danger to health and safety;
(c) information on risks to health and safety—
 (i) identified by the risk assessment under regulation 3 of the Management Regulations, or
 (ii) arising out of the conduct of another contractor's undertaking and of which the contractor in control of the worker ought reasonably to be aware; and
(d) any other information necessary to enable the worker to comply with the relevant statutory provisions.

(10) A contractor must not begin work on a construction site unless reasonable steps have been taken to prevent access by unauthorised persons to that site.

(11) A contractor must ensure, so far as is reasonably practicable, that the requirements of Schedule 2 are complied with so far as they affect the contractor or any worker under that contractor's control.

Guidance 15

Who is a contractor?

147 Anyone who directly employs or engages construction workers or manages construction is a contractor. Contractors include sub-contractors, any individual, sole trader, self-employed worker, or business that carries out, manages or controls construction work as part of their business. This also includes companies that use their own workforce to do construction work on their own premises. The duties on contractors apply whether the workers under their control are employees, self-employed or agency workers.

148 Where contractors are involved in design work, including for temporary works, they also have duties as designers (see paragraphs 79–93).

Why is a contractor important?

149 Contractors and the workers under their control are those most at risk of injury and ill health. They can influence the way work is carried out to secure their own health and safety and that of others affected. They have an important role in planning, managing and monitoring the work (in liaison with the principal contractor, where appropriate) to ensure risks are properly controlled. The key to this is the

Guidance 15

proper coordination of the work, underpinned by good communication and cooperation with others involved.

What must a contractor do?

150 Contractors have a number of specific duties. They must also comply with the requirements of regulation 8 as they apply to contractors (see paragraphs 57–71). These include the requirements:

(a) on anyone appointing a designer or contractor (such as the contractor appointing a sub-contractor) to ensure the designer or contractor has the skills, knowledge and experience and, where relevant, organisational capability to carry out the work for which they are being appointed; and

(b) to cooperate with other dutyholders.

Making clients aware of their duties

151 Contractors must not carry out any construction work on a project unless they are satisfied that the client is aware of the duties the client has under CDM 2015. In cases where the contractor is the only one involved, they must liaise directly with the client to establish this. Liaison can be done as part of routine business during early meetings with the client to discuss the project. Contractors should make sure they have a sufficient knowledge of client duties as they affect the project so they can give proper advice. The level of advice will depend on the knowledge and experience of the client and the complexities of the project.

Planning, managing and monitoring construction work
General

152 Contractors are required to plan, manage and monitor the construction work under their control so it is carried out in a way that controls the risks to health and safety. The effort devoted to planning, managing and monitoring should be proportionate to the size and complexity of the project and the nature of risks involved.

153 On projects involving more than one contractor, this will involve the contractor coordinating the planning, management and monitoring of their own work with that of the principal contractor and other contractors, and where appropriate the principal designer. Such coordination could involve regular progress meetings with other dutyholders to ensure that the contractor's arrangements for planning, managing and monitoring their own work can feed into, and remain consistent with, the project-wide arrangements. For single contractor projects, the arrangements to plan, manage and monitor the construction phase will normally be simpler. Paragraphs 154–159 provide guidance in each circumstance.

Planning

154 In planning the work, the contractor must take into account the risks to those who may be affected, eg members of the public and those carrying out the construction work. Planning should cover the same considerations as those for the principal contractor (see paragraphs 123–124), including considering the risks and ensuring the measures needed to protect those affected are in place.

155 On projects involving more than one contractor, each contractor must plan their own work so it is consistent with the project-wide arrangements. Contractors should expect help from other dutyholders, eg the client who must provide the pre-construction information (Appendix 2 gives more guidance on the provision of pre-construction information).

Guidance 15

156 On single contractor projects, the contractor is responsible for planning the construction phase and for drawing up the construction phase plan before setting up the construction site. The client must provide any relevant pre-construction information they possess and the time and other resources to help the contractor do this. See paragraph 161 and Appendix 3 for further guidance on drawing up the construction phase plan.

Managing
157 The arrangements for managing construction work must take into account the same issues that principal contractors must consider (see paragraphs 125–128).

Monitoring
158 The contractor should monitor their work to ensure that the health and safety precautions are appropriate, remain in place and are followed in practice. Effective monitoring by the contractor must address the same issues principal contractors must consider (see paragraph 129). This includes using a mix of measures to check performance and taking prompt action when issues arise.

159 On projects involving more than one contractor, as part of the duty to cooperate with other dutyholders, the contractor should provide the principal contractor with any relevant information that stems from their own monitoring so the principal contractor can monitor the management of health and safety at a project-wide level.

Complying with directions and the construction phase plan
160 For projects involving more than one contractor, the contractor is required to comply with any directions to secure health and safety given to them by the principal designer or principal contractor. They are also required to comply with the parts of the construction phase plan that are relevant to their work, including the site rules (see Appendix 3).

Drawing up a construction phase plan
161 For single contractor projects, the contractor must ensure a construction phase plan is drawn up as soon as practicable **before** the construction site is set up. Guidance on contractors' duties in relation to the construction phase plan is set out in Appendix 3. There is more guidance, including a template for a construction phase plan, for contractors working on small-scale, routine and domestic projects in the HSE leaflet *Construction phase plan (CDM 2015): What you need to know as a busy builder.*[4] *CDM wizard* is a smartphone app template for a construction phase plan produced by the Construction Industry Training Board (CITB).[5]

Appointing and employing workers
Appointing workers
162 When a contractor employs or appoints an individual to work on a construction site, they should make enquiries to make sure the individual:

(a) has the skills, knowledge, training and experience to carry out the work they will be employed to do in a way that secures health and safety for anyone working on the site; or
(b) is in the process of obtaining them. Paragraphs 163–173 give guidance on what a contractor should consider when appointing anyone who has gaps in the skills, knowledge or experience necessary for the work.

163 Sole reliance should not be placed on industry certification cards or similar being presented to them as evidence that a worker has the right qualities. Nationally recognised qualifications (such as National Vocational Qualifications (NVQs) and Scottish Vocational Qualifications (SVQs)) can provide contractors with

Guidance 15

assurance that the holder has the skills, knowledge, training and experience to carry out the task(s) for which they are appointed. Contractors should recognise that training on its own is not enough. Newly trained individuals need to be supervised and given the opportunity to gain positive experience of working in a range of conditions.

164 When appointing individuals who may be skilled but who do not have any formal qualifications, contractors may need to assess them in the working environment.

Training workers
165 To establish whether training is necessary for any worker, a contractor should:

(a) assess the existing health and safety skills, knowledge, training and experience of their workers;
(b) compare these existing attributes with the range of skills, knowledge, training and experience they will need for the job; and
(c) identify any shortfall between (a) and (b). The difference between the two will be the 'necessary training'.

As a general rule, if the person being assessed demonstrates the required qualities, no further training should be needed.

166 This assessment should take account of the training required by other health and safety legislation (eg section 2 of HSWA) as well as that needed to meet the requirements of CDM 2015.

167 Assessing training needs should be an ongoing process throughout the project. Further training may be required if:

(a) the risks to which people are exposed alter due to a change in their working tasks;
(b) new technology or equipment is introduced; or
(c) the system of work changes.

Skills can also decline if they are not used regularly. Particular attention should be paid to people who deputise for others on an occasional basis – they may need more frequent further training than those who do the work regularly.

168 Contractors should also consider 'softer skills', such as the ability to foresee risk, maintain sensitivity to risk, anticipate mistakes others might make and to communicate clearly, as well as the more technical skills workers require for their work.

Providing supervision
169 A contractor who employs workers or manages workers under their control must ensure that appropriate supervision is provided. The level of supervision provided will depend on the risks to health and safety involved, and the skills, knowledge, training and experience of the workers concerned.

170 Workers will require closer supervision if they are young, inexperienced, or starting a new work activity. Other factors that should be considered when assessing the level of supervision needed include the level of individuals' safety awareness, education, physical agility, literacy and attitude. Even experienced workers may need an appropriate level of supervision if they do not have some or all of the skills, knowledge, training and experience required for the job and the risks involved. Workers should always know how to get supervisory help, even when a supervisor is not present.

Guidance 15

171 Supervisors are a vital part of effective management arrangements. Effective supervisors are those who have the skills, knowledge, training, experience and leadership qualities to suit the job in hand. Good communication and people management skills on site are important qualities for supervisors. Where site workers are promoted to a supervisory role, they should be provided with nationally recognised site supervisor training which includes leadership and communication skills.

172 The role of the supervisor may include team leading, briefing and carrying out toolbox talks. It may also include coaching and encouragement of individual workers and supporting other formal and informal means of engaging with workers. The supervisor has a particularly important part to play as a front-line decision maker in emergencies or when workers on site face immediate risks that may require work to stop (see paragraph 68).

Providing information and instructions

173 Contractors should provide their employees and workers under their control with the information and instructions they need to carry out their work without risk to health and safety. This must include:

(a) suitable site induction where this has not been provided by the principal contractor. In such cases, the guidance provided in paragraphs 133–134 for principal contractors is relevant to contractors;

(b) the procedures to be followed in the event of serious and imminent danger to health and safety. These should make clear that any worker exposed to any such danger should stop work immediately, report it to the contractor and go to a place of safety. The procedures should:
   (i) include details of the person to whom such instances should be reported and who has the authority to take whatever prompt action is needed;
   (ii) take account of the relevant requirements which set out provisions relating to emergency procedures, emergency routes and exits and fire detection and fire-fighting;

(c) information on the hazards on site relevant to their work (eg site traffic), the risks associated with those hazards and the control measures put in place (eg the arrangements for managing site traffic).

See paragraphs 69–71 for more guidance.

Preventing unauthorised access to the site

174 A contractor must not begin work on a construction site unless reasonable steps have been taken to prevent unauthorised access to the site. For projects involving more than one contractor that are:

(a) small and straightforward, this can be carried out via a phone call or at an early meeting with the principal contractor (who is required to ensure reasonable steps are taken in this respect) before the contractor starts work on site;

(b) larger and more complex (eg where different contractors are authorised to access different parts of the site), contractors should liaise with the principal contractor to make sure they understand which parts of the site they are authorised to access and when before they start work.

175 For projects involving only one contractor, the contractor must do whatever is proportionate to prevent unauthorised access before starting work on the site. In these circumstances, the guidance provided for principal contractors is also relevant for contractors (see paragraphs 135–136).

Guidance 15

Providing welfare facilities

176 Contractors are required to provide welfare facilities which meet the minimum requirements set out in Schedule 2. This duty only extends to the provision of welfare facilities for the contractor's own employees who are working on a construction site or anyone else working under their control. Facilities must be made available before any construction work starts and should be maintained until the end of the project.

177 The duty is as *far as is reasonably practicable*, so contractors should do whatever is proportionate in providing the welfare facilities set out in Schedule 2. Guidance on what is proportionate in providing welfare facilities on construction sites is contained in HSE's Construction Information Sheet, *Provision of welfare facilities during construction work*.[6]

178 On projects involving more than one contractor, meeting this duty will involve discussing and agreeing with the principal contractor who has a similar duty to provide welfare facilities (see paragraphs 137–138). For projects involving only one contractor, the contractor themselves must ensure that suitable welfare facilities are available.

Working for domestic clients

179 A contractor's role when working on a project for a domestic client is no different to their role when carrying out work for a commercial client. They must still carry out the duties set out in regulations 8 and 15 in proportion to the risks involved in the project. But, the effect of regulation 7 is to transfer the duties of the domestic client to another dutyholder, including the contractor in certain circumstances. See Appendix 6 for guidance on how this affects what contractors must do on domestic projects.

PART 4 General requirements for all construction sites

Guidance

180 **Part 4** sets out a number of provisions that only relate to work carried out on the construction site. See HSE's construction web pages at www.hse.gov.uk/construction for guidance on these provisions.

Regulation 16 Application of Part 4

Regulation 16

(1) This part applies only to a construction site.

(2) A contractor carrying out construction work must comply with the requirements of this Part so far as they affect the contractor or any worker under the control of the contractor or relate to matters within the contractor's control.

(3) A domestic client who controls the way in which any construction work is carried out by a person at work must comply with the requirements of this Part so far as they relate to matters within the client's control.

Regulation 17 Safe places of construction work

Regulation 17

(1) There must, so far as is reasonably practicable, be suitable and sufficient safe access and egress from—

(a) every construction site to every other place provided for the use of any person whilst at work; and
(b) every place construction work is being carried out to every other place to which workers have access within a construction site.

(2) A construction site must be, so far as is reasonably practicable, made and kept safe for, and without risks to, the health of any person at work there.

(3) Action must be taken to ensure, so far as is reasonably practicable, that no person uses access to or egress from or gains access to any construction site which does not comply with the requirements of paragraph (1) or (2).

(4) A construction site must, so far as is reasonably practicable, have sufficient working space and be arranged so that it is suitable for any person who is working or who is likely to work there, taking account of any necessary work equipment likely to be used there.

Regulation 18 Good order and site security

Regulation 18

(1) Each part of a construction site must, so far as is reasonably practicable, be kept in good order and those parts in which construction work is being carried out must be kept in a reasonable state of cleanliness.

Regulation 18

(2) Where necessary in the interests of health and safety, a construction site must, so far as is reasonably practicable, and in accordance with the level of risk posed, comply with either or both of the following—

(a) have its perimeter identified by suitable signs and be arranged so that its extent is readily identifiable; or
(b) be fenced off.

(3) No timber or other material with projecting nails (or similar sharp object) must—

(a) be used in any construction work; or
(b) be allowed to remain in any place,

if the nails (or similar sharp object) may be a source of danger to any person.

Regulation 19 Stability of structures

Regulation 19

(1) All practicable steps must be taken, where necessary to prevent danger to any person, to ensure that any new or existing structure does not collapse if, due to the carrying out of construction work, it—

(a) may become unstable; or
(b) is in a temporary state of weakness or instability.

(2) Any buttress, temporary support or temporary structure must—

(a) be of such design and installed and maintained so as to withstand any foreseeable loads which may be imposed on it; and
(b) only be used for the purposes for which it was designed, and installed and is maintained.

(3) A structure must not be so loaded as to render it unsafe to any person.

Regulation 20 Demolition or dismantling

Regulation 20

(1) The demolition or dismantling of a structure must be planned and carried out in such a manner as to prevent danger or, where it is not practicable to prevent it, to reduce danger to as low a level as is reasonably practicable.

(2) The arrangements for carrying out such demolition or dismantling must be recorded in writing before the demolition or dismantling work begins.

Regulation 21 Explosives

Regulation 21

(1) So far as is reasonably practicable, explosives must be stored, transported and used safely and securely.

(2) An explosive charge may be used or fired only if suitable and sufficient steps have been taken to ensure that no person is exposed to risk of injury from the explosion or from projected or flying material caused by the explosion.

Regulation 22 Excavations

Regulation 22

(1) All practicable steps must be taken to prevent danger to any person, including, where necessary, the provision of supports or battering, to ensure that—

(a) no excavation or part of an excavation collapses;
(b) no material forming the walls or roof of, or adjacent to, any excavation is dislodged or falls; and
(c) no person is buried or trapped in an excavation by material which is dislodged or falls.

(2) Suitable and sufficient steps must be taken to prevent any person, work equipment, or any accumulation of material from falling into any excavation.

(3) Suitable and sufficient steps must be taken, where necessary, to prevent any part of an excavation or ground adjacent to it from being overloaded by work equipment or material.

(4) Construction work must not be carried out in an excavation where any supports or battering have been provided in accordance with paragraph (1) unless—

(a) the excavation and any work equipment and materials which may affect its safety have been inspected by a competent person—
 (i) at the start of the shift in which the work is to be carried out;
 (ii) after any event likely to have affected the strength or stability of the excavation; and
 (iii) after any material unintentionally falls or is dislodged; and
(b) the person who carried out the inspection is satisfied that construction work can be safely carried out there.

(5) Where the person carrying out an inspection informs the person on whose behalf the inspection is carried out of any matter about which they are not satisfied (under regulation 24(1)), construction work must not be carried out in the excavation until the matter has been satisfactorily remedied.

Regulation 23 Cofferdams and caissons

Regulation 23

(1) A cofferdam or caisson must be—

(a) of suitable design and construction;
(b) appropriately equipped so that workers can gain shelter or escape if water or materials enter it; and
(c) properly maintained.

(2) A cofferdam or caisson must not be used to carry out construction work unless—

(a) the cofferdam or caisson and any work equipment and materials which may affect its safety have been inspected by a competent person—
 (i) at the start of the shift in which the work is to be carried out; and
 (ii) after any event likely to have affected the strength or stability of the cofferdam or caisson; and
(b) the person who carried out the inspection is satisfied that construction work can be safely carried out there.

Regulation 23

(3) Where the person carrying out an inspection informs the person on whose behalf the inspection is carried out of any matter about which they are not satisfied (under regulation 24(1)), construction work must not be carried out in the cofferdam or caisson until the matter has been satisfactorily remedied.

Regulation 24 Reports of inspections

Regulation 24

(1) Where a person who carries out an inspection under regulation 22 or 23 is not satisfied that construction work can be carried out safely at the place inspected, that person must—

(a) inform the person on whose behalf the inspection was carried out, before the end of the shift within which the inspection is completed, of the matters that could give rise to a risk to the safety of any person; and
(b) prepare a report which must include—
 (i) the name and address of the person on whose behalf the inspection was carried out;
 (ii) the location of the place of construction work inspected;
 (iii) a description of the place of construction work or part of that place inspected (including any work equipment and materials);
 (iv) the date and time of the inspection;
 (v) details of any matter identified that could give rise to a risk to the safety of any person;
 (vi) details of any action taken as a result of any matter identified in sub-paragraph (v);
 (vii) details of any further action considered necessary; and
 (viii) the name and position of the person making the report; and
(c) provide the report, or a copy of it, to the person on whose behalf the inspection was carried out, within 24 hours of completing the inspection to which the report relates.

(2) Where the person who carries out an inspection works under the control of another (whether as an employee or otherwise) the person in control must ensure the person who carries out the inspection complies with the requirements of paragraph (1).

(3) The person on whose behalf the inspection was carried out must—

(a) keep the report or a copy of it available for inspection by an inspector for the Executive—
 (i) at the site where the inspection was carried out until the construction work is completed; and
 (ii) after that for 3 months; and
(b) send to the inspector such extracts from or copies of the report as the inspector may from time to time require.

(4) This regulation does not require the preparation of more than one report where more than one inspection is carried out under regulation 22(4)(a)(i) or 23(2)(a)(i) within a 7 day period.

Regulation 25 Energy distribution installations

(1) Where necessary to prevent danger, energy distribution installations must be suitably located, periodically checked and clearly indicated.

(2) Where there is a risk to construction work from overhead electric power cables—

(a) they must be directed away from the area of risk; or
(b) the power must be isolated and, where necessary, earthed.

(3) If it is not reasonably practicable to comply with paragraph (2)(a) or (b), suitable warning notices must be provided together with one or more of the following—

(a) barriers suitable for excluding work equipment which is not needed;
(b) suspended protections where vehicles need to pass beneath the cables; or
(c) measures providing an equivalent level of safety.

(4) Construction work which is liable to create a risk to health or safety from an underground service, or from damage to or disturbance of it, must not be carried out unless suitable and sufficient steps (including any steps required by this regulation) have been taken to prevent the risk, so far as is reasonably practicable.

Regulation 26 Prevention of drowning

(1) Where, in the course of construction work, a person is at risk of falling into water or other liquid with a risk of drowning, suitable and sufficient steps must be taken to—

(a) prevent, so far as is reasonably practicable, the person falling;
(b) minimise the risk of drowning in the event of a fall; and
(c) ensure that suitable rescue equipment is provided, maintained and, when necessary, used so that a person may be promptly rescued in the event of a fall.

(2) Suitable and sufficient steps must be taken to ensure the safe transport of any person conveyed by water to or from a place of work.

(3) Any vessel used to convey any person by water to or from a place of work must not be overcrowded or overloaded.

Regulation 27 Traffic routes

(1) A construction site must be organised in such a way that, so far as is reasonably practicable, pedestrians and vehicles can move without risks to health or safety.

(2) Traffic routes must be suitable for the persons or vehicles using them, sufficient in number, in suitable positions and of sufficient size.

(3) A traffic route does not satisfy paragraph (2) unless suitable and sufficient steps are taken to ensure that—

Regulation 27

(a) pedestrians or vehicles may use it without causing danger to the health or safety of persons near it;
(b) any door or gate for pedestrians which leads onto a traffic route is sufficiently separated from that traffic route to enable pedestrians to see any approaching vehicle or plant from a place of safety;
(c) there is sufficient separation between vehicles and pedestrians to ensure safety or, where this is not reasonably practicable—
  (i) other means for the protection of pedestrians are provided, and
  (ii) effective arrangements are used for warning any person liable to be crushed or trapped by any vehicle of its approach;
(d) any loading bay has at least one exit for the exclusive use of pedestrians; and
(e) where it is unsafe for pedestrians to use a gate intended primarily for vehicles, at least one door for pedestrians is provided in the immediate vicinity of the gate, is clearly marked and is kept free from obstruction.

(4) Each traffic route must be—

(a) indicated by suitable signs where necessary for reasons of health or safety;
(b) regularly checked; and
(c) properly maintained.

(5) No vehicle is to be driven on a traffic route unless, so far as is reasonably practicable, that traffic route is free from obstruction and permits sufficient clearance.

Regulation 28 Vehicles

Regulation 28

(1) Suitable and sufficient steps must be taken to prevent or control the unintended movement of any vehicle.

(2) Where a person may be endangered by the movement of a vehicle, suitable and sufficient steps to give warning to any person who is liable to be at risk from the movement of the vehicle must be taken by either or both—

(a) the driver or operator of the vehicle, or
(b) where another person is directing the driver or operator because, due to the nature of the vehicle or task, the driver or operator does not have full visibility, the person providing directions.

(3) A vehicle being used for the purposes of construction work must, when being driven, operated or towed be—

(a) driven, operated or towed in such a manner as is safe in the circumstances; and
(b) loaded in such a way that it can be driven, operated or towed safely.

(4) A person must not ride, or be required or permitted to ride, on any vehicle being used for the purposes of construction work otherwise than in a safe place in that vehicle provided for that purpose.

(5) A person must not remain, or be required or permitted to remain, on any vehicle during the loading or unloading of any loose material unless a safe place of work is provided and maintained for that person.

(6) Suitable and sufficient measures must be taken to prevent a vehicle from falling into any excavation or pit, or into water, or overrunning the edge of any embankment or earthwork.

Regulation 29 Prevention of risk from fire, flooding or asphyxiation

Regulation 29

(1) Suitable and sufficient steps must be taken to prevent, so far as is reasonably practicable, the risk of injury to a person during the carrying out of construction work arising from—

(a) fire or explosion;
(b) flooding; or
(c) any substance liable to cause asphyxiation.

Regulation 30 Emergency procedures

Regulation 30

(1) Where necessary in the interests of the health or safety of a person on a construction site, suitable and sufficient arrangements for dealing with any foreseeable emergency must be made and, where necessary, implemented, and those arrangements must include procedures for any necessary evacuation of the site or any part of it.

(2) In making arrangements under paragraph (1), account must be taken of—

(a) the type of work for which the construction site is being used;
(b) the characteristics and size of the construction site and the number and location of places of work on that site;
(c) the work equipment being used;
(d) the number of persons likely to be present on the site at any one time; and
(e) the physical and chemical properties of any substances or materials on, or likely to be on, the site.

(3) Where arrangements are made under paragraph (1), suitable and sufficient steps must be taken to ensure that—

(a) each person to whom the arrangements extend is familiar with those arrangements; and
(b) the arrangements are tested by being put into effect at suitable intervals.

Regulation 31 Emergency routes and exits

Regulation 31

(1) Where necessary in the interests of the health or safety of a person on a construction site, a sufficient number of suitable emergency routes and exits must be provided to enable any person to reach a place of safety quickly in the event of danger.

(2) The matters in regulation 30(2) must be taken into account when making provision under paragraph (1).

(3) An emergency route or exit must lead as directly as possible to an identified safe area.

(4) An emergency route or exit and any traffic route giving access to it must be kept clear and free from obstruction and, where necessary, provided with emergency lighting so that it may be used at any time.

(5) Each emergency route or exit must be indicated by suitable signs.

Regulation 32 Fire detection and fire-fighting

Regulation 32

(1) Where necessary in the interests of the health or safety of a person on a construction site, suitable and sufficient fire-fighting equipment and fire detection and alarm systems must be provided and located in suitable places.

(2) The matters in regulation 30(2) must be taken into account when making provision under paragraph (1).

(3) Fire-fighting equipment or fire detection and alarm systems must be examined and tested at suitable intervals and properly maintained.

(4) Fire-fighting equipment which is not designed to come into use automatically must be easily accessible.

(5) Each person at work on a construction site must, so far as is reasonably practicable, be instructed in the correct use of fire-fighting equipment which it may be necessary for the person to use.

(6) Where a work activity may give rise to a particular risk of fire, a person must not carry out work unless suitably instructed.

(7) Fire-fighting equipment must be indicated by suitable signs.

Regulation 33 Fresh air

Regulation 33

(1) Suitable and sufficient steps must be taken to ensure, so far as is reasonably practicable, that each construction site, or approach to a construction site, has sufficient fresh or purified air to ensure that the site or approach is safe and without risks to health or safety.

(2) Any plant used for the purpose of complying with paragraph (1) must, where necessary for reasons of health and safety, include an effective device to give visible or audible warning of any failure of the plant.

Regulation 34 Temperature and weather protection

Regulation 34

(1) Suitable and sufficient steps must be taken to ensure, so far as reasonably practicable, that during working hours the temperature at a construction site that is indoors is reasonable having regard to the purpose for which that place is used.

(2) Where necessary to ensure the health or safety of persons at work on a construction site that is outdoors, the construction site must, so far as is reasonably practicable, be arranged to provide protection from adverse weather, having regard to—
    (a) the purpose for which the site is used; and
    (b) any protective clothing or work equipment provided for the use of any person at work there.

Regulation 35 Lighting

Regulation 35

(1) Each construction site and approach and traffic route to that site must be provided with suitable and sufficient lighting, which must be, so far as is reasonably practicable, by natural light.

(2) The colour of any artificial lighting provided must not adversely affect or change the perception of any sign or signal provided for the purposes of health or safety.

(3) Suitable and sufficient secondary lighting must be provided in any place where there would be a risk to the health or safety of a person in the event of the failure of primary artificial lighting.

PART 5 General

Regulation 36 Enforcement in respect of fire

Regulation 36

The enforcing authority for regulations 30 and 31 (so far as those regulations relate to fire) and regulation 32, in respect of a construction site which is contained within or forms part of premises occupied by persons other than those carrying out construction work, or any activity related to this work, is—

(a) in England and Wales, the enforcing authority within the meaning of article 25 of the Regulatory Reform (Fire Safety) Order 2005 in respect of premises to which that Order applies; or

(b) in Scotland, the enforcing authority within the meaning of section 61 of the Fire (Scotland) Act 2005 in respect of premises to which Part 3 of that Act applies.

Regulation 37 Transitional and saving provisions

Regulation 37

Schedule 4, which makes transitional and saving provisions, has effect.

Guidance 37

181 The provisions in Schedule 4 recognise that there will be projects that start before CDM 2015 comes into force on 6 April 2015 and continue beyond that date. For these projects, the following transitional arrangements apply.

Appointments

182 Under CDM 2007, a client was required to appoint a CDM co-ordinator to provide advice and assistance to the client; to make arrangements for the coordination and implementation of health and safety measures during the pre-construction phase and to identify and collect the pre-construction information. CDM 2015 removes this role and replaces it with the principal designer role.

183 Schedule 4 provides transitional arrangements for projects which span 6 April 2015. For projects involving more than one contractor which started before 6 April 2015, where by that date the client **has not** appointed a CDM co-ordinator, the client:

(a) **must** appoint a principal designer, as soon as practicable, if the construction phase **has not** started;

(b) is **not required** to appoint a principal designer if the construction phase **has** started, but may do so if they wish. If they choose **not** to appoint a principal designer, the principal contractor takes on the responsibility for the health and safety file (see Appendix 4). In these circumstances, any designer involved with the project should provide information about any residual risks in designs to the principal contractor.

Where on 6 April 2015 the client **has** appointed a CDM co-ordinator, they must appoint a principal designer within six months – ie by 6 October 2015.

Guidance 37

184 The CDM co-ordinator must then comply with the duties listed in paragraph 5 of Schedule 4 for the duration of their appointment. These duties broadly reflect the duties of a CDM co-ordinator under CDM 2007, but also reflect the arrangements under CDM 2015 relating to the construction phase plan (see Appendix 3) and the health and safety file (see Appendix 4).

185 During the transitional period, CDM co-ordinators do not have to satisfy the criteria for a principal designer under regulation 5(1)(a) that they should be a designer with control over the pre-construction phase of the project, nor do they have to comply with the principal designer duties under regulation 11.

Other transitional provisions

186 These are as follows:

(a) any pre-construction information, construction phase plan or health and safety file provided in accordance with the requirements of CDM 2007 are recognised as meeting the requirements of the equivalent provisions in CDM 2015;

(b) notification of a project in accordance with CDM 2007 is recognised as a notification for the purposes of CDM 2015; and

(c) a principal contractor appointed under CDM 2007 will be considered to be a principal contractor for the purposes of CDM 2015.

Regulation 38 Revocation and consequential amendments

Regulation 38

(1) The 2007 Regulations are revoked.

(2) The amendments in Schedule 5 have effect.

Regulation 39 Review

Regulation 39

(1) The Secretary of State must from time to time—

(a) carry out a review of regulations 1 to 36;
(b) set out the conclusions of the review in a report; and
(c) publish the report.

(2) In carrying out the review the Secretary of State must, so far as is reasonable, have regard to how Council Directive 92/57/EEC on the implementation of minimum safety and health requirements at temporary or mobile construction sites (which is implemented by means of regulations 1 to 36), is implemented in other Member States.

(3) The report must in particular—

(a) set out the objectives intended to be achieved by the regulatory system established by those regulations;
(b) assess the extent to which those objectives are achieved; and
(c) assess whether those objectives remain appropriate and, if so, the extent to which they could be achieved with a system that imposes less regulation.

| **Regulation 39** | *(4) The first report under this regulation must be published before the end of the period of five years beginning with the day on which these Regulations come into force.*

(5) Reports under this regulation are afterwards to be published at intervals not exceeding five years. |
|---|---|

SCHEDULE 1 Particulars to be notified under regulation 6

Schedule 1	Regulation 6
	1. The date of forwarding the notice.
	2. The address of the construction site or precise description of its location.
	3. The name of the local authority where the construction site is located.
	4. A brief description of the project and the construction work that it entails.
	5. The following contact details of the client: name, address, telephone number and (if available) an email address.
	6. The following contact details of the principal designer: name, address, telephone number and (if available) an email address.
	7. The following contact details of the principal contractor: name, address, telephone number and (if available) an email address.
	8. The date planned for the start of the construction phase.
	9. The time allocated by the client under regulation 4(1) for the construction work.
	10. The planned duration of the construction phase.
	11. The estimated maximum number of people at work on the construction site.
	12. The planned number of contractors on the construction site.
	13. The name and address of any contractor already appointed.
	14. The name and address of any designer already appointed.
	15. A declaration signed by or on behalf of the client that the client is aware of the client duties under these Regulations.

SCHEDULE 2 Minimum welfare facilities required for construction sites

Schedule 2

Regulation 4(2)(b), 13(4)(c) and 15(11)

Sanitary conveniences

1.— *(1) Suitable and sufficient sanitary conveniences must be provided or made available at readily accessible places.*

(2) So far as is reasonably practicable, rooms containing sanitary conveniences must be adequately ventilated and lit.

(3) So far as is reasonably practicable, sanitary conveniences and the rooms containing them must be kept in a clean and orderly condition.

(4) Separate rooms containing sanitary conveniences must be provided for men and women, except where and so far as each convenience is in a separate room, the door of which is capable of being secured from the inside.

Washing facilities

2.— *(1) Suitable and sufficient washing facilities, including showers if required by the nature of the work or for health reasons, must, so far as is reasonably practicable, be provided or made available at readily accessible places.*

(2) Washing facilities must be provided—

(a) in the immediate vicinity of every sanitary convenience, whether or not also provided elsewhere; and
(b) in the vicinity of any changing rooms required by paragraph 4, whether or not provided elsewhere.

(3) Washing facilities must include—

(a) a supply of clean hot and cold, or warm, water (which must be running water so far as is reasonably practicable);
(b) soap or other suitable means of cleaning; and
(c) towels or other suitable means of drying.

(4) Rooms containing washing facilities must be sufficiently ventilated and lit.

(5) Washing facilities and the rooms containing them must be kept in a clean and orderly condition.

(6) Subject to paragraph (7), separate washing facilities must be provided for men and women, except where they are provided in a room the door of which is capable of being secured from inside and the facilities in each room are intended to be used by only one person at a time.

Schedule 2

(7) *Sub-paragraph (6) does not apply to facilities which are provided for washing hands, forearms and the face only.*

Drinking water

3.— *(1) An adequate supply of wholesome drinking water must be provided or made available at readily accessible and suitable places.*

(2) Where necessary for reasons of health or safety, every supply of drinking water must be conspicuously marked by an appropriate sign.

(3) Where a supply of drinking water is provided, a sufficient number of suitable cups or other drinking vessels must also be provided, unless the supply of drinking water is in a jet from which persons can drink easily.

Changing rooms and lockers

4.— *(1) Suitable and sufficient changing rooms must be provided or made available at readily accessible places if a worker—*

(a) *has to wear special clothing for the purposes of construction work; and*
(b) *cannot, for reasons of health or propriety, be expected to change elsewhere.*

(2) Where necessary for reasons of propriety, there must be separate changing rooms for, or separate use of rooms by, men and women.

(3) Changing rooms must—

(a) *be provided with seating; and*
(b) *include, where necessary, facilities to enable a person to dry any special clothing and any personal clothing or effects.*

(4) Suitable and sufficient facilities must, where necessary, be provided or made available at readily accessible places to enable persons to lock away—

(a) *any special clothing which is not taken home;*
(b) *their own clothing which is not worn during working hours; and*
(c) *their personal effects.*

Facilities for rest

5.— *(1) Suitable and sufficient rest rooms or rest areas must be provided or made available at readily accessible places.*

(2) Rest rooms and rest areas must—

(a) *be equipped with an adequate number of tables and adequate seating with backs for the number of persons at work likely to use them at any one time;*
(b) *where necessary, include suitable facilities for any woman at work who is pregnant or who is a nursing mother to rest lying down;*

Schedule 2

(c) include suitable arrangements to ensure that meals can be prepared and eaten;
(d) include the means for boiling water; and
(e) be maintained at an appropriate temperature.

SCHEDULE 3 Work involving particular risks

Schedule 3

Regulation 12(2)

1. Work which puts workers at risk of burial under earthfalls, engulfment in swampland or falling from a height, where the risk is particularly aggravated by the nature of the work or processes used or by the environment at the place of work or site.

2. Work which puts workers at risk from chemical or biological substances constituting a particular danger to the safety or health of workers or involving a legal requirement for health monitoring.

3. Work with ionizing radiation requiring the designation of controlled or supervised areas under regulation 16 of the Ionising Radiations Regulations 1999.

4. Work near high voltage power lines.

5. Work exposing workers to the risk of drowning.

6. Work on wells, underground earthworks and tunnels.

7. Work carried out by divers having a system of air supply.

8. Work carried out by workers in caissons with a compressed air atmosphere.

9. Work involving the use of explosives.

10. Work involving the assembly or dismantling of heavy prefabricated components.

SCHEDULE 4 Transitional and saving provisions

Schedule 4

Regulation 37

1. *In this Schedule—*

"CDM co-ordinator" means a person appointed under regulation 14(1) of the 2007 Regulations;

"competent" means competent to perform any requirement and avoid contravening any prohibition imposed on a person by or under any of the relevant statutory provisions;

"relevant project" means a project which began before 6th April 2015.

2. *These Regulations apply to a relevant project with the modifications specified in this Schedule.*

Projects with no existing CDM co-ordinator or principal contractor

3.— *(1) This paragraph modifies the application of these Regulations in relation to a relevant project where, immediately before 6th April 2015—*

 (a) no CDM co-ordinator or principal contractor is appointed for the project under the 2007 Regulations;
 (b) there is more than one contractor, or it is reasonably foreseeable that more than one contractor will be working on a project; and
 (c) the construction phase has started.

 (2) Regulation 5 does not apply to the project.

 (3) The client may appoint in writing a designer as principal designer.

 (4) The client must appoint in writing a contractor as principal contractor as soon as is practicable after 6th April 2015.

 (5) The principal contractor must draw up a construction phase plan or make arrangements for a construction phase plan to be drawn up under regulation 12(1) and (2) as soon as is practicable after 6th April 2015 and the requirement that the plan must be drawn up during the pre-construction phase and before setting up a construction site is disapplied.

 (6) The client, other than a domestic client, must ensure that the principal contractor complies with sub-paragraph (5) and the client duty in regulation 4(5)(a) is disapplied.

Schedule 4

(7) If the client does not appoint a principal designer, the principal contractor must—

(a) prepare a health and safety file under regulation 12(5) as soon as is practicable after 6th April 2015 and the requirement for the file to be prepared during the pre-construction phase is disapplied; and
(b) ensure that the health and safety file is reviewed, updated and revised from time to time under regulation 12(6).

(8) If the client does not appoint a principal designer—

(a) the references to the principal designer in regulations 4(5)(b) and 9(3)(b) are treated as references to the principal contractor;
(b) the client duty in regulation 4(6)(a) does not apply; and
(c) the principal contractor duties in regulations 12(7) and 13(5) do not apply.

(9) Where a client, other than a domestic client, fails to appoint a principal contractor under sub-paragraph (4) the client must fulfil the duties of the principal contractor specified in these Regulations, as modified by this paragraph.

(10) Where the client is a domestic client—

(a) regulation 7(2) does not apply; and
(b) if the client fails to appoint a principal contractor under sub-paragraph (4) the principal contractor for the project is the contractor in control of the construction phase.

Projects with an existing CDM co-ordinator

4.— (1) This paragraph and paragraphs 5 and 6 apply where, immediately before 6th April 2015, there is a CDM co-ordinator appointed for a relevant project.

(2) Where this paragraph applies, the appointment of the CDM co-ordinator continues to have effect for the purposes of these Regulations until a principal designer is appointed or the project comes to an end.

(3) The client must appoint in writing a principal designer for the project before the 6th October 2015 unless the project comes to an end on or before that date.

(4) Where the appointment of a CDM co-ordinator continues to have effect under sub-paragraph (2)—

(a) the CDM co-ordinator must comply with the duties in paragraph 5;
(b) the duties in regulations 9(3)(b) and 12(7) to provide information to the principal designer are treated as duties to provide information to the CDM co-ordinator; and
(c) the duty in regulation 13(5) to liaise with the principal designer is treated as a duty to liaise with the CDM co-ordinator.

Duties of CDM co-ordinator during transitional period

5.— (1) The CDM co-ordinator must—

(a) cooperate with any other person working on or in relation to a project at the same or an adjoining construction site, to the extent necessary to

Schedule 4

      enable any person with a duty or function under these Regulations to fulfil that duty or function;

(b)   *where the CDM co-ordinator works under the control of another, report to that person anything they are aware of in relation to the project which is likely to endanger their own health or safety or that of others;*

(c)   *ensure that suitable arrangements are made and implemented for the coordination of health and safety measures during the planning and preparation for the construction phase, including facilitating—*
   - (i)   *cooperation and coordination between all persons working on the pre-construction phase of the project; and*
   - (ii)   *the application of the general principles of prevention;*

(d)   *liaise with the principal contractor over—*
   - (i)   *the content of the health and safety file;*
   - (ii)   *the information which the principal contractor needs to prepare the construction phase plan; and*
   - (iii)   *any design development which may affect planning and management of the construction work;*

(e)   *where no or partial pre-construction information has been supplied to the CDM co-ordinator by the client under regulation 10 of the 2007 Regulations, assist the client to comply with regulation 4(4) of these Regulations;*

(f)   *unless the information has already been provided under regulation 20(2)(b) of the 2007 Regulations, provide any pre-construction information that is in the possession or control of the CDM co-ordinator, promptly and in a convenient form, to every designer and contractor appointed, or being considered for appointment, to the project;*

(g)   *take all reasonable steps to ensure that designers comply with their duties under regulation 9 of these Regulations;*

(h)   *take all reasonable steps to ensure cooperation between designers and the principal contractor during the construction phase in relation to any design or change to a design;*

(i)   *if a health and safety file has not been prepared under regulation 20(2)(e) of the 2007 Regulations, prepare a health and safety file that complies with the requirements of regulation 12(5) of these Regulations;*

(j)   *review, update and revise the health and safety file from time to time to take account of the work and any changes that have occurred;*

(k)   *if the CDM co-ordinator's appointment continues to have effect immediately before the project ends, pass the health and safety file to the client at the end of the project;*

(l)   *if a principal designer is appointed, pass the health and safety file and all other relevant health and safety information in the CDM co-ordinator's possession to the principal designer, as soon as is practicable after the appointment.*

(2)   *The CDM co-ordinator must not arrange for or instruct a worker to carry out or manage design or construction work unless the worker is competent or under the supervision of a competent person.*

Duties disapplied pending appointment of principal designer

6.— (1)   *The duties in regulation 5(1)(a) and (3) do not apply to a project referred to in paragraph 4(1).*

(2)   *The following duties do not apply to a project referred to in paragraph 4(1) until the principal designer is appointed—*

(a)   *the duties in regulations 4(5)(b) and (6)(a);*

Schedule 4

(b) the duties of the principal designer in regulations 11 and 12(3), (5), (6), (8) and (10).

(3) If a client fails to make the appointment required by paragraph 4(3) the client must fulfil the duties of a principal designer in regulations 11 and 12 on and after 6th October 2015.

Projects with only one contractor

7. Where a relevant project has only one contractor and the construction phase has started, the contractor must draw up a construction phase plan, or make arrangements for a construction phase plan to be drawn up, under regulation 15(5) and (6) as soon as is practicable after 6th April 2015 and the requirement that the plan must be drawn up prior to setting up a construction site is disapplied.

Savings

8.— (1) Where, immediately before 6th April 2015 there is a principal contractor appointed for a relevant project under regulation 14(2) of the 2007 Regulations, for the purposes of these Regulations that principal contractor is treated on and after 6th April 2015 as having been appointed under regulation 5(1)(b) of these Regulations.

(2) For the purposes of these Regulations, on and after the 6th April 2015—

(a) a health and safety file prepared for a relevant project under regulation 20(2)(e) of the 2007 Regulations is treated as a health and safety file prepared under regulation 12(5) of these Regulations;
(b) a construction phase plan drawn up for a relevant project under regulation 23 of the 2007 Regulations is treated as a construction phase plan drawn up under regulation 12(1) or 15(5) of these Regulations;
(c) pre-construction information provided for a relevant project under regulation 10 of the 2007 Regulations is treated as pre-construction information provided under regulation 4(4) of these Regulations;
(d) notice given for a relevant project under regulation 21 of the 2007 Regulations is treated as notice given under regulation 6 of these Regulations.

SCHEDULE 5 Amendments

Schedule 5

Regulation 38

Description of instrument	Reference	Extent of amendment
Factories Act 1961	*1961 c. 34*	*In section 176(1) in the definition of "building operation" and "work of engineering construction" for "2007" (SI 2007/320) substitute "2015".*
Workplace (Health Safety and Welfare) Regulations 1992	*SI 1992/3004*	*In regulation 3(1)(b) for "2007" substitute "2015".*
Work in Compressed Air Regulations 1996	*SI 1996/1656*	*In regulation 2(1) in the definition of "the 2007 Regulations" substitute "2015" in both places it appears.* *In regulation 3(1) for "2007" substitute "2015" and for "2(3)" substitute "6(1)".* *In regulation 5(3) for "2007" substitute "2015".* *In regulation 13(2)(a) for "39, 40 and 44(3) of the 2007" substitute "30, 31 and 35(3) of the 2015".* *In regulation 13(2)(d) for "39(1) of the 2007" substitute "30(1) of the 2015".* *In regulation 14(1) for "41 of the 2007" substitute "32 of the 2015".*
Railway Safety (Miscellaneous Provisions) Regulations 1997	*SI 1997/553*	*In regulation 2(1) in the definition of "construction work" for "2007" substitute "2015".*

Schedule 5	Health and Safety (Enforcing Authority) Regulations 1998	SI 1998/494	In regulation 2(1) in the definition of "construction work" and "contractor" for "2007" substitute "2015". In regulation 2A(5)(a) for "2007" substitute "2015". In Schedule 2, paragraph 4(a)(i) for "2(3)" substitute "6(1)" and for "2007" substitute "2015"
	Provision and Use of Work Equipment Regulations 1998	SI 1998/2306	In regulation 6(5)(e) for "31(4) or 32(2)" substitute "22(4) or 23(2)" and for "2007" substitute "2015".
	Gas Safety (Installation and Use) Regulations 1998	SI 1998/2451	In regulation 2(4)(d) for "2007" substitute "2015".
	Work at Height Regulations 2005	SI 2005/735	In regulation 2(1) in the definition of "construction work" for "2007" substitute "2015".
	Regulatory Reform (Fire Safety) Order 2005	SI 2005/1541	In article 25(2)(a) for "2007" substitute "2015" and for "46(1)" substitute "36".
	Health and Safety (Enforcing Authority for Railways and other Guided Transport Systems) Regulations 2006	SI 2006/557	In regulation 2(1) in the definitions of "construction work" and "contractor" for "2007" substitute "2015".
	REACH Enforcement Regulations 2008	SI 2008/2852	In paragraph 1(d)(i)(aa) of Part 3 of Schedule 3 for "2(3)" substitute "6(1)" and for "2007" substitute "2015".
	Reporting of Injuries, Diseases and Dangerous Occurrences Regulations 2013	SI 2013/1471	In regulation 2(1) in the definition of "construction site" for "2007" substitute "2015"

Appendix 1 The general principles of prevention

1 These principles are a requirement of the Management Regulations and apply to all industries, including construction. They provide a framework to identify and implement measures to control risks on a construction project.

2 The general principles of prevention are to:

(a) avoid risks;
(b) evaluate the risks which cannot be avoided;
(c) combat the risks at source;
(d) adapt the work to the individual, especially regarding the design of workplaces, the choice of work equipment and the choice of working and production methods, with a view, in particular, to alleviating monotonous work, work at a predetermined work rate and to reducing their effect on health;
(e) adapt to technical progress;
(f) replace the dangerous by the non-dangerous or the less dangerous;
(g) develop a coherent overall prevention policy which covers technology, organisation of work, working conditions, social relationships and the influence of factors relating to the working environment;
(h) give collective protective measures priority over individual protective measures; and
(i) give appropriate instructions to employees.

Appendix 2 Pre-construction information

1 This Appendix gives guidance on the requirements for pre-construction information and the actions on each dutyholder. Appendix 5 shows how pre-construction information relates to and influences other types of information during a construction project involving more than one contractor.

What is pre-construction information?

2 Pre-construction information provides the health and safety information needed by:

(a) designers and contractors who are bidding for work on the project, or who have already been appointed to enable them to carry out their duties;
(b) principal designers and principal contractors in planning, managing, monitoring and coordinating the work of the project.

Pre-construction information provides a basis for the preparation of the construction phase plan (see Appendix 3). Some material may also be relevant to the preparation of the health and safety file (see Appendix 4).

3 Pre-construction information is defined as information about the project that is already in the client's possession or which is reasonably obtainable by or on behalf of the client. The information must:

(a) **be relevant to the particular project;**
(b) **have an appropriate level of detail; and**
(c) **be proportionate to the risks involved.**

4 Pre-construction information should be gathered and added to as the design process progresses and reflect new information about the health and safety risks and how they should be managed. Preliminary information gathered at the start of the project is unlikely to be sufficient.

5 When pre-construction information is complete, it must include proportionate information about:

(a) the project, such as the client brief and key dates of the construction phase;
(b) the planning and management of the project such as the resources and time being allocated to each stage of the project and the arrangements to ensure there is cooperation between dutyholders and the work is coordinated;
(c) the health and safety hazards of the site, including design and construction hazards and how they will be addressed;
(d) any relevant information in an existing health and safety file.

6 The information should be in a convenient form and be clear, concise and easily understandable to help other dutyholders involved in the project to carry out their duties.

What must dutyholders do?

The client
7 The client has the main duty for providing pre-construction information. They must provide this information as soon as practicable to each:

(a)   designer (including the principal designer); and
(b)   contractor (including the principal contractor)

being considered for appointment, or already appointed to the project. For projects involving more than one contractor, the client can expect help from the principal designer appointed for the project (see paragraphs 11–13 of this Appendix) who must assist the client in drawing this information together and providing it to the designers and contractors involved. For single contractor projects, it is the client's responsibility alone – although they should liaise with the contractor (and any designer) they appoint to provide whatever information is needed.

8 The pre-construction information will evolve as the project progresses towards the construction phase. At first, drawing together the information should involve identifying relevant documents the client already holds. These might include a health and safety file produced as a result of earlier construction work, any surveys or assessments that have already been carried out (eg asbestos surveys), structural drawings etc. For projects involving more than one contractor, the client must pass this information to the principal designer as soon after their appointment as possible. In liaison with the principal designer, the client should then:

(a)   assess the adequacy of this information to see if there are significant gaps;
(b)   take reasonable steps to obtain the information needed to fill any gaps identified by, eg commissioning relevant surveys; and
(c)   then provide the information to every designer and contractor as soon as practicable.

9 The stage at which it is practicable to provide information will depend on a number of factors such as the scale and complexity of the project, when dutyholders are appointed and when information is obtained. However, the client, together with the principal designer, must also take account of when designers and contractors will need pre-construction information to enable them to carry out their duties. For example:

(a)   designers or contractors who are seeking appointment for work on the project should have sufficient information made available to them at a time which allows them to put together a bid based on a clear understanding of the nature of the work involved;
(b)   designers already appointed should be provided with sufficient information at a stage early enough to enable them to judge whether it is reasonably practicable to eliminate any foreseeable health and safety risks in the design process and, where it is not, the steps they should take to reduce or control the remaining risks. It may not be possible to provide this information all at once, in which case it should be provided as soon as it becomes available;
(c)   contractors already appointed should be provided with the information they will need to plan, manage and monitor their work.

The designer
10 The designer must take account of the pre-construction information when preparing or modifying designs. They must be provided with this information by the client as soon as practicable (see paragraphs 7–9 of this Appendix), assisted by the principal designer where appropriate (see paragraphs 11–13 of this Appendix). The information should be:

(a) sufficient to enable the designer to judge whether it is reasonably practicable to eliminate foreseeable risks in their designs, and, where it is not, help identify the steps they should take to reduce and control the remaining risks; and

(b) provided at a stage where designers can take account of it – as early in the design process as is practicable.

The principal designer

11 The principal designer must help the client in providing the pre-construction information to each designer and contractor appointed, or being considered for appointment. The extent of the help required will depend on the nature of the project, the risks involved and the client's level of knowledge and experience of construction work. Taking this into account, the principal designer should agree with the client the level of support the client needs to ensure the information is made available when others need it.

12 Soon after their appointment, the principal designer should be provided with any relevant information the client already holds. This might include any health and safety file produced as a result of earlier construction work, any surveys that have already been carried out (eg asbestos surveys), structural drawings etc. The principal designer must then help the client to:

(a) assess the adequacy of this information to see if there are significant gaps; and

(b) take reasonable steps to obtain the information needed to fill the gaps identified by, eg commissioning surveys.

13 As far as it is within their control, the principal designer must then work with the client to provide the information in a convenient form and as soon as practicable. The information provided to those seeking appointment must be sufficient and in good time to allow them to put together a bid based on a clear understanding of the nature of the work involved. After their appointment, the stage at which information is provided will depend on factors such as the scale and complexity of the project, and when the information is obtained. However, the principal designer, together with the client, must also take account of when designers and contractors will need pre-construction information to enable them to carry out their duties. The client guidance (see paragraph 9 of this Appendix) is also relevant for principal designers.

The principal contractor

14 The principal contractor has no specific duty in relation to pre-construction information. However, they must liaise with the principal designer for the duration of the principal designer's appointment and share any information relevant to the planning, management, monitoring or coordination of the pre-construction phase.

The contractor

15 The contractor has no specific duty in relation to pre-construction information. However, for projects involving more than one contractor, contractors must cooperate with the client, principal designer and principal contractor to ensure the pre-construction information is right.

Appendix 3 The construction phase plan

1 This Appendix gives guidance on the requirements for the construction phase plan and the actions on each dutyholder. Appendix 5 shows how the construction phase plan relates to and influences other types of information during a construction project involving more than one contractor.

What is a construction phase plan?

2 A construction phase plan is a document that must record the:

(a) health and safety arrangements for the construction phase;
(b) site rules; and
(c) where relevant, specific measures concerning work that falls within one or more of the categories listed in Schedule 3.

3 The plan must record the arrangements for managing the significant health and safety risks associated with the construction phase of a project. It is the basis for communicating these arrangements to all those involved in the construction phase, so it should be easy to understand and as simple as possible.

4 In considering what information is included, the emphasis is that it:

(a) **is relevant to the project;**
(b) **has sufficient detail to clearly set out the arrangements, site rules and special measures needed to manage the construction phase; but**
(c) **is still proportionate to the scale and complexity of the project and the risks involved.**

The plan should **not** include documents that get in the way of a clear understanding of what is needed to manage the construction phase, such as generic risk assessments, records of how decisions were reached or detailed safety method statements.

5 The following list of topics should be considered when drawing up the plan:

(a) a description of the project such as key dates and details of key members of the project team;
(b) the management of the work including:
    (i) the health and safety aims for the project;
    (ii) the site rules;
    (iii) arrangements to ensure cooperation between project team members and coordination of their work, eg regular site meetings;
    (iv) arrangements for involving workers;
    (v) site induction;
    (vi) welfare facilities; and
    (vii) fire and emergency procedures;

(c) the control of any of the specific site risks listed in Schedule 3 where they are relevant to the work involved.

What must dutyholders do?

The client
6 The client must ensure a construction phase plan is drawn up **before** the construction phase begins. For projects involving more than one contractor, the principal contractor is responsible for drawing up the plan or for making arrangements for it to be drawn up (see paragraphs 11–14 of this Appendix). For single contractor projects, it is the contractor who is responsible for ensuring that the plan is drawn up (see paragraphs 15–17 of this Appendix).

7 The client must ensure that the principal contractor (or, where relevant, the contractor) is provided with all the available relevant information they need to draw up the plan, eg the pre-construction information (see Appendix 2).

8 The client must also ensure that:

(a) when it is drawn up, the plan adequately addresses the arrangements for managing the risks; and
(b) the principal contractor (or contractor) regularly reviews and revises the plan to ensure it takes account of any changes that occur as construction progresses and continues to be fit for purpose.

The designer
9 The designer has no specific duty in relation to the construction phase plan. However, the designer must take all reasonable steps to provide with the design sufficient information about aspects of the design to help contractors (including principal contractors) to comply with their duties. This should include information about the significant risks designers have been unable to eliminate through the design process and the steps designers have taken to reduce or control those risks. They must continue to cooperate with contractors and principal contractors as the construction phase progresses to ensure that they are kept up to date with any design changes.

The principal designer
10 The principal designer must help the principal contractor to prepare the construction phase plan by providing any relevant information they hold. This includes:

(a) the pre-construction information given to them by the client and which they have an important role in pulling together and providing (see Appendix 2); and
(b) any information given to them by designers about the risks that have not been eliminated through the design process and the steps taken to reduce or control those risks.

Before the start of the construction phase, the principal designer should regularly check that the principal contractor has the information needed to prepare the plan. They must continue to liaise with the principal contractor as the construction phase progresses to share any information relevant to the planning and management of the construction phase.

The principal contractor
11 For projects involving more than one contractor, the principal contractor must take the lead in preparing, reviewing, updating and revising the construction phase

plan. They must draw up the plan or make arrangements for it to be drawn up during the pre-construction phase and **before** the construction site is set up.

12 The principal contractor should expect help from both the client and principal designer in doing this. The client's duty is to ensure that the plan is drawn up and the principal designer's duty is to help the principal contractor by providing any relevant information they hold (see paragraph 10 of this Appendix). This information should include:

(a) the pre-construction information that the client must provide to every designer and contractor involved in the project and which the principal designer will have been involved in preparing; and
(b) any information provided by designers about the risks that designers have been unable to eliminate through the design process and the steps they have taken to reduce or control them.

The principal contractor must also liaise with the contractors to ensure that the plan takes into account their views on the arrangements for managing the construction phase.

13 Where the plan includes site rules, the rules should cover (but not be limited to) topics such as personal protective equipment, parking, use of radios and mobile phones, smoking, restricted areas, hot works and emergency arrangements. The rules should be clear and easily understandable. They should be brought to the attention of everyone on site who should be expected to follow them. The principal contractor should also consider any special requirements, eg it might be necessary to have translations of the site rules available.

14 The principal contractor must ensure that the construction phase plan is appropriately reviewed, updated and revised from time to time. The plan is a working document and will need to be reviewed regularly enough to address significant changes to the risks involved in the work or in the effectiveness of the controls that have been put in place. This means that the principal contractor must monitor how effective the plan is in addressing identified risks and whether it is being implemented properly. Ensuring the plan remains fit for purpose must also involve co-operating with:

(a) the contractors who are most likely to see if the arrangements for controlling health and safety risks are working; and
(b) the principal designer and designers when changes in designs during the construction phase have implications for the plan.

The contractor

15 For projects involving more than one contractor, the contractor must follow the parts of the construction phase plan prepared by the principal contractor that are relevant to their work. The contractor should also liaise with the principal contractor to pass on their views on the effectiveness of the plan in managing the risks.

16 For single contractor projects, the contractor has the responsibility for ensuring that a construction phase plan is drawn up. They must either draw up a plan themselves, or make arrangements for it to be drawn up, as soon as practicable **before** setting up the construction site. In preparing the plan they must cooperate with the client and any designers involved in the project and take account of sources of relevant information such as the pre-construction information (see Appendix 2).

17 Further guidance, including a template for a construction phase plan,[5] is provided for contractors working on small scale, routine and domestic projects on

HSE's website. *CDM wizard* is a smartphone app template for a construction phase plan produced by CITB.[6]

Appendix 4 The health and safety file

1 This Appendix gives guidance on the preparation, provision and retention of a health and safety file and the actions on each dutyholder. Appendix 5 shows how the health and safety file relates to and influences other types of information during a construction project involving more than one contractor.

What is the health and safety file?

2 The health and safety file is defined as a file appropriate to the characteristics of the project, containing relevant health and safety information to be taken into account during any subsequent project. **The file is only required for projects involving more than one contractor.**

3 The file must contain information about the current project likely to be needed to ensure health and safety during any subsequent work, such as maintenance, cleaning, refurbishment or demolition. When preparing the health and safety file, information on the following should be considered for inclusion:

(a) a brief description of the work carried out;
(b) any hazards that have not been eliminated through the design and construction processes, and how they have been addressed (eg surveys or other information concerning asbestos or contaminated land);
(c) key structural principles (eg bracing, sources of substantial stored energy – including pre- or post-tensioned members) and safe working loads for floors and roofs;
(d) hazardous materials used (eg lead paints and special coatings);
(e) information regarding the removal or dismantling of installed plant and equipment (eg any special arrangements for lifting such equipment);
(f) health and safety information about equipment provided for cleaning or maintaining the structure;
(g) the nature, location and markings of significant services, including underground cables; gas supply equipment; fire-fighting services etc;
(h) information and as-built drawings of the building, its plant and equipment (eg the means of safe access to and from service voids and fire doors).

4 There should be enough detail to allow the likely risks to be identified and addressed by those carrying out the work. However, the level of detail should be proportionate to the risks. The file should **not** include things that will be of no help when planning future construction work such as pre-construction information, the construction phase plan, contractual documents, safety method statements etc. Information must be in a convenient form, clear, concise and easily understandable.

What must dutyholders do?

The client

5 The client must ensure that the principal designer prepares the health and safety file for a project. As the project progresses, the client must ensure that the principal designer regularly updates, reviews and revises the health and safety file to take account of the work and any changes that have occurred. The client should be aware that if the principal designer's appointment finishes before the end of the project, the principal designer must pass the health and safety file to the principal contractor, who then must take on the responsibility for the file.

6 Once the project is finished, the client should expect the principal designer to pass them the health and safety file. In cases where the principal designer has left the project before it finishes, it will be for the principal contractor to pass the file to the client.

7 The client must then retain the file and ensure it is available to anyone who may need it for as long as it is relevant – normally the lifetime of the building – to enable them to comply with health and safety requirements during any subsequent project. It can be kept electronically, on paper, on film, or any other durable form.

8 If a client disposes of their interest in the building, they must give the file to the individual or organisation who takes on the client duties and ensure that the new client is aware of the nature and purpose of the file. If they sell part of a building, any relevant information in the file must be passed or copied to the new owner. If the client leases out all or part of the building, arrangements should be made for the file to be made available to leaseholders. If the leaseholder acts as a client for a future construction project, the leaseholder and the original client must arrange for the file to be made available to the new principal designer.

The designer

9 Where it is not possible to eliminate health and safety risks when preparing or modifying designs, designers must ensure appropriate information is included in the health and safety file about the reasonably practicable steps they have taken to reduce or control those risks. This will involve liaising with:

(a) the principal designer, in helping them carry out their duty to prepare, update, review and revise the health and safety file. This should continue for as long as the principal designer's appointment on the project lasts; or
(b) the principal contractor, where design work is carried out after the principal designer's appointment has finished and where changes need to be made to the health and safety file. In these circumstances, it will be the principal contractor's duty to make those changes, but the designer must ensure that the principal contractor has the appropriate information to update the file.

This information should be provided to the principal designer and principal contractor as early as possible before the designer's work ends on the project.

The principal designer

10 The principal designer must prepare the health and safety file. They are accountable to the client and should liaise closely to agree the structure and content of the file as soon as practicable after appointment. In preparing the file, the principal designer should expect the client to provide any health and safety file that may exist from an earlier project.

11 The principal designer must also cooperate with the rest of the project team and should expect their cooperation in return. Cooperation with the principal

contractor is particularly important in agreeing the structure and content of the information included in the file. Liaison with designers and other contractors is also important. They may hold information that is useful for the health and safety file, which may be difficult to obtain after they have left the project.

12 The principal designer, in cooperation with other members of the project team, must also ensure that the file is appropriately updated, reviewed and revised to ensure it takes account of any changes that occur as the project progresses.

13 The principal designer must pass the updated file to the client at the end of the project. In doing this, they should ensure the client understands the structure and content of the file and its significance for any subsequent project. If the principal designer's appointment finishes before the end of the project, they must pass the file to the principal contractor who must then take on responsibility for it. In doing this, the principal designer should ensure the principal contractor is aware of any outstanding issues that may need to be taken into account when reviewing, updating and revising the file.

The principal contractor
14 For the duration of the principal designer's appointment, the principal contractor plays a secondary role in ensuring the health and safety file is fit for purpose. They must provide the principal designer with any relevant information that needs to be included in the health and safety file.

15 Where the principal designer's appointment finishes before the end of the project, the principal contractor must take on responsibility for ensuring that the file is reviewed, updated and revised for the remainder of the project. At the end of the project the principal contractor must pass the file to the client. In doing this, they should ensure the client understands the structure and content of the file and its significance for any subsequent project.

The contractor
16 The contractor has no specific duties placed on them in relation to the health and safety file.

Appendix 5 How different types of information relate to and influence each other in a construction project involving more than one contractor: A summary

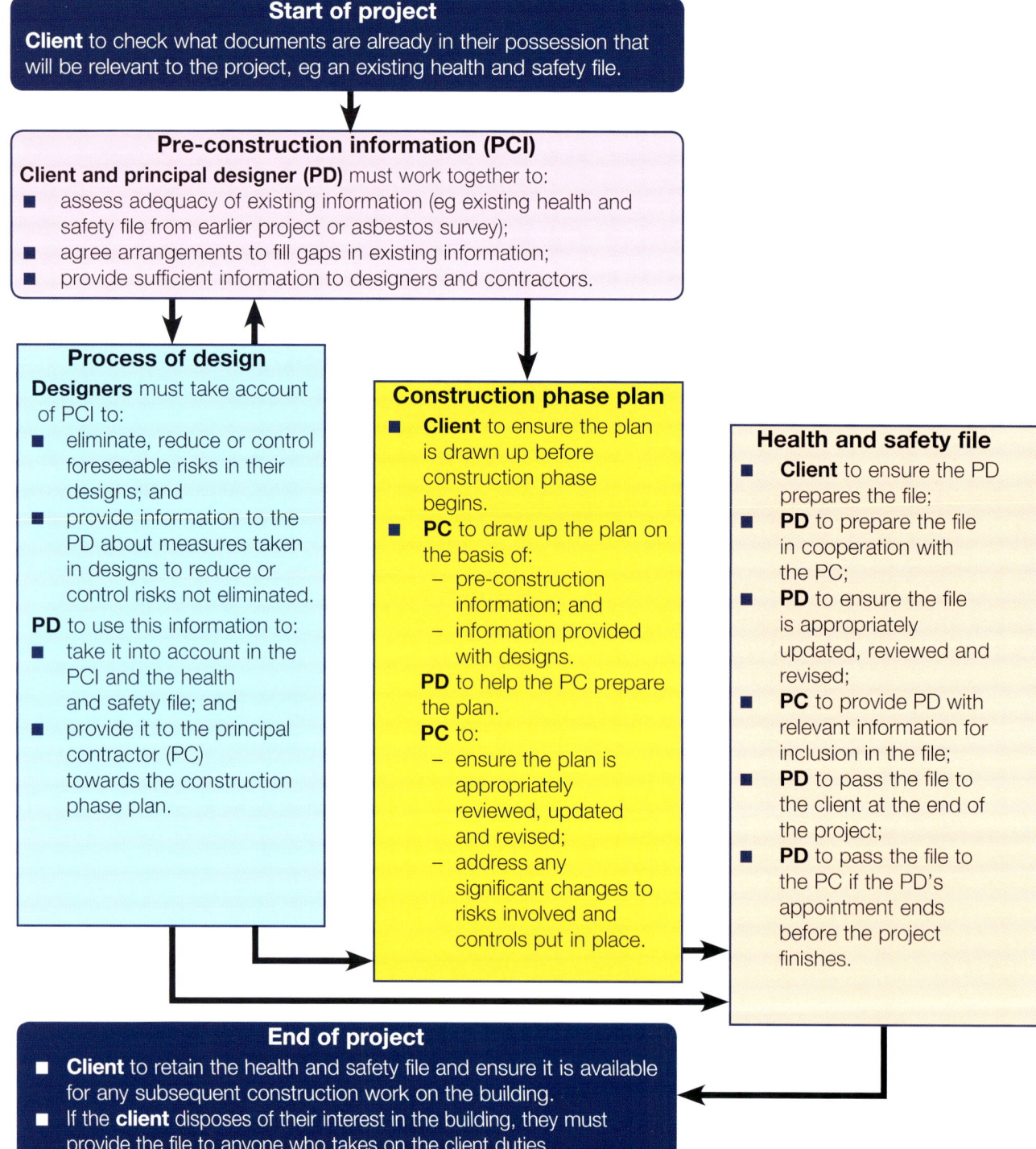

Note: This diagram shows how the various types of information relate to each other and influence the content of other types of information during the construction process (the arrows show the possible different flows of information). So, for example as pre-construction information is developed, this influences the risks designers should consider and the information they provide about how their designs reduce or control foreseeable risks. In turn, this may influence further development of the pre-construction information, as well as the construction phase plan and the health and safety file.

Appendix 6 Working for a domestic client

1 The role of designers, principal designers, principal contractors and contractors when working on a project for a domestic client, is normally no different to their role when working for a commercial client. They have the same duties and should carry them out in the same way as they would for a commercial client. However, the effect of regulation 7 is to transfer the client duties to other dutyholders when working for domestic clients.

2 Guidance for domestic clients in relation to CDM 2015 is set out in paragraphs 53–56. The following paragraphs set out what other dutyholders need to do as a result. Figure 1 is a flow chart showing the transfer of client duties from a domestic client to other dutyholders involved.

Domestic projects involving only one contractor

3 On these projects the client's duties are transferred to the contractor and they must carry out the client's duties as well as their own. In practice, this should involve contractors doing no more than they have done in the past to comply with health and safety legislation. **Compliance with their own duties as a contractor will be taken as compliance with the relevant client duties**, to the extent necessary given the risks involved in the project.

4 As a result of the contractor taking on the client duties, any designers involved in the project will work to the contractor in their role as the 'client'.

Domestic projects involving more than one contractor

Transfer of the client duties to the principal contractor
5 On these projects, it will normally be the principal contractor who takes on the client duties and they will need to comply with these duties as well as their own. If the domestic client does not appoint a principal contractor, the role of principal contractor falls to the contractor in control of the construction phase of the project.

6 As a result of a principal contractor taking on the client duties, the principal designer involved in the project will work to the principal contractor in their role as the 'client'. If the domestic client does not appoint a principal designer, the role of the principal designer falls to the designer in control of the pre-construction phase of the project.

Transfer of client duties to principal designer
7 Domestic clients can choose to have a written agreement with the principal designer to transfer the client duties to the principal designer. In this case, the principal designer must fulfil the duties of the client as well as their own and the principal contractor will work to the principal designer as the 'client'.

Figure 1 How CDM 2015 applies to domestic clients

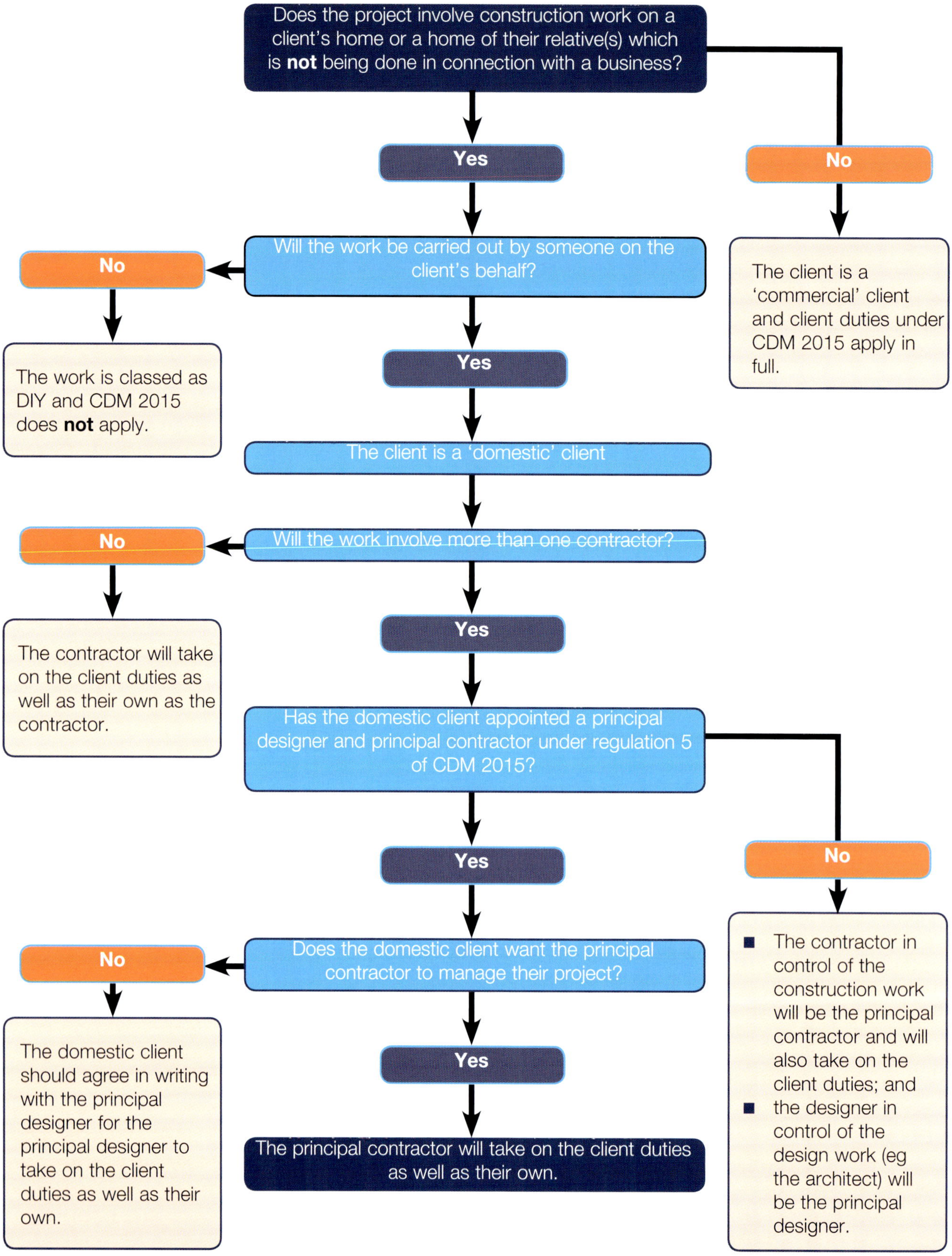

- **Does the project involve construction work on a client's home or a home of their relative(s) which is not being done in connection with a business?**
  - **No** → The client is a 'commercial' client and client duties under CDM 2015 apply in full.
  - **Yes** ↓
- **Will the work be carried out by someone on the client's behalf?**
  - **No** → The work is classed as DIY and CDM 2015 does **not** apply.
  - **Yes** ↓
- **The client is a 'domestic' client**
- **Will the work involve more than one contractor?**
  - **No** → The contractor will take on the client duties as well as their own as the contractor.
  - **Yes** ↓
- **Has the domestic client appointed a principal designer and principal contractor under regulation 5 of CDM 2015?**
  - **No** →
    - The contractor in control of the construction work will be the principal contractor and will also take on the client duties; and
    - the designer in control of the design work (eg the architect) will be the principal designer.
  - **Yes** ↓
- **Does the domestic client want the principal contractor to manage their project?**
  - **No** → The domestic client should agree in writing with the principal designer for the principal designer to take on the client duties as well as their own.
  - **Yes** → The principal contractor will take on the client duties as well as their own.

References and further reading

References

1 *Consulting employees on health and safety: A brief guide to the law* Leaflet INDG232(rev2) HSE Books 2013 www.hse.gov.uk/pubns/indg232.htm

2 PAS 91:2013 *Construction related procurement. Prequalification questionnaires* British Standards Institution
www.shop.bsigroup.com/Navigate-by/PAS/PAS-91-2013/

3 *Workplace health, safety and welfare. Workplace (Health, Safety and Welfare) Regulations 1992. Approved Code of Practice and guidance* L24 (Second edition) HSE Books 2013 ISBN 978 0 7176 6583 9 www.hse.gov.uk/pubns/books/l24.htm

4 *Construction phase plan (CDM 2015): What you need to know as a busy builder* Construction Information Sheet CIS80 HSE Books 2015
www.hse.gov.uk/pubns/cis80.pdf

5 *CDM wizard* CITB smartphone app www.citb.co.uk/cdmregs

6 *Provision of welfare facilities during construction work* Construction Information Sheet CIS59 HSE Books 2010 www.hse.gov.uk/pubns/cis59.htm

Further reading

Health and safety in construction HSG150 (Third edition) HSE Books 2006
ISBN 978 0 7176 6182 4 www.hse.gov.uk/pubns/books/hsg150.htm

The absolutely essential health and safety toolkit for the smaller construction contractor Leaflet INDG344(rev2) HSE Books www.hse.gov.uk/pubns/indg344.htm

Want building work done safely? A quick guide for clients on the Construction (Design and Management) Regulations 2015 Leaflet INDG411(rev1)
HSE Books 2015 www.hse.gov.uk/pubns/indg411.htm

Glossary of acronyms and terms

ACOP Approved Code of Practice.

AOGBO 2013 Application to Areas Outside Great Britain Order 2013.

CDM 2007 Construction (Design and Management) Regulations 2007.

CDM 2015 Construction (Design and Management) Regulations 2015.

CITB Construction Industry Training Board.

HSE Health and Safety Executive.

HSWA Health and Safety at Work etc Act 1974.

LWIT Leadership and Worker Involvement Toolkit.

The Management Regulations Management of Health and Safety at Work Regulations 1999.

must 'must' is used only where there is an absolute duty, ie an explicit legal requirement to take a certain action which is not qualified by terms such as 'so far as reasonably practicable'.

ONR Office for Nuclear Regulation.

ORR Office of Rail Regulation.

PAS 91 Publicly Available Specification 91.

reasonably practicable balancing the level of risk against the measures needed to control the real risk in terms of money, time or trouble. However, you do not need to take action if it would be grossly disproportionate to the level of risk (see www.gov.uk/risk/faq.htm for the most up-to-date explanation of what 'reasonably practicable' means.)

renewable energy zone areas outside the territorial sea designated for the exploration or exploitation of energy from water or winds (see section 84 of the Energy Act 2004).

should 'should' is used to indicate what to do to comply with legal requirements which are qualified by terms such as 'so far as reasonably practicable'.

significant risks not necessarily those that involve the greatest risks, but those (including health risks) that are not likely to be obvious, are unusual, or likely to be difficult to manage effectively.

SSIP Safety Schemes in Procurement.

territorial sea means the belt of waters extending from the coast of Great Britain up to 12 nautical miles.

Workplace Regulations Workplace (Health, Safety and Welfare) Regulations 1992.

Further information

For information about health and safety visit https://books.hse.gov.uk or http://www.hse.gov.uk. You can view HSE guidance online and order priced publications from the website. HSE priced publications are also available from bookshops.

To report inconsistencies or inaccuracies in this guidance email: commissioning@williamslea.com.

British Standards can be obtained in PDF or hard copy formats from BSI: http://shop.bsigroup.com or by contacting BSI Customer Services for hard copies only Tel: 0846 086 9001 email: cservices@bsigroup.com.

The Stationery Office publications are available from The Stationery Office, PO Box 29, Norwich NR3 1GN Tel: 0333 202 5070 Fax: 0333 202 5080. E-mail:customer.services@tso.co.uk Website: www.tso.co.uk. They are also available from bookshops.

Statutory Instruments can be viewed free of charge at www.legislation.gov.uk where you can also search for changes to legislation.